Primer

Student Workbook

1-888-854-MATH (6284) - mathusee.com
sales@mathusee.com

Primer Student Workbook: Introduction to Math

©2012 Math-U-See, Inc.
Published and distributed by Demme Learning

mathusee.com

1-888-854-6284 or +1 717-283-1448 | demmelearning.com
Lancaster, Pennsylvania USA

ISBN 978-1-60826-065-2
Revision Code 1018

Printed in the United States of America by Bindery Associates LLC
 2 3 4 5 6 7 8 9 10

For information regarding CPSIA on this printed material call: 1-888-854-6284
and provide reference #1018-101518

APPLICATION AND ENRICHMENT PAGES

This edition of the *Primer Student Workbook* includes extra activity pages titled "Application and Enrichment." You will find one enrichment page after the last Systematic Review page for each lesson. These activities are intended to do the following:

- Provide enjoyable practice of lesson concepts
- Stimulate thinking by presenting concepts in different formats
- Include activities suitable for a wide range of learning styles
- Enrich learning with additional age-appropriate activities

The Application and Enrichment pages may be scheduled any time after the student has completed the corresponding lesson. Students may need a little help getting started with some of them. Our hope is that students will finish *Primer* with smiles on their faces and a positive attitude towards math, so keep your approach lighthearted and fun.

Check your instruction manual for more games and teaching tips. There are no solution pages for the *Primer* Application and Enrichment pages.

Count the blocks and circle the correct number.

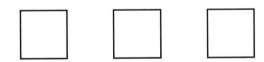

0 1 2 3 4 5 6 7 8 9

. . . .

0 1 2 3 4 5 6 7 8 9

0 1 2 3 4 5 6 7 8 9

LESSON PRACTICE

Count the blocks and circle the correct number.

0 1 2 3 4 5 6 7 8 9

0 1 2 3 4 5 6 7 8 9

PRIMER LESSON PRACTICE 1B 7

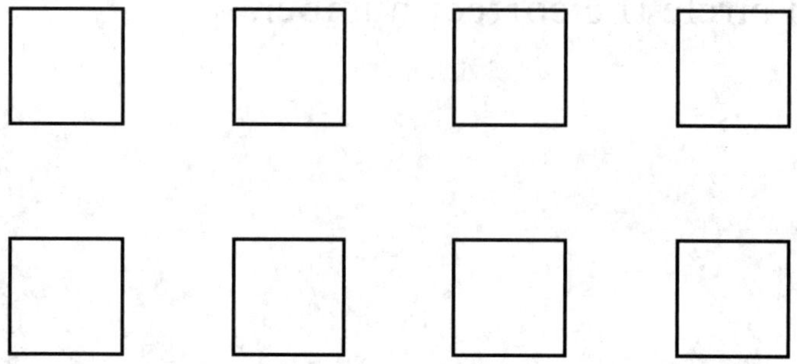

□ □ □ □

□ □ □ □

0 1 2 3 4 5 6 7 8 9

LESSON PRACTICE

Count the blocks and circle the correct number.

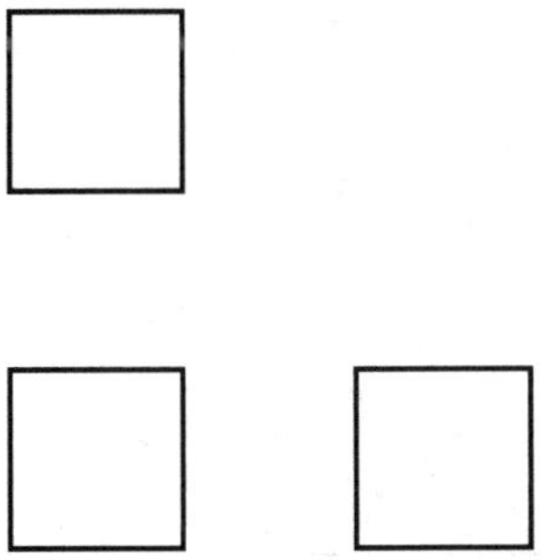

0 1 2 3 4 5 6 7 8 9

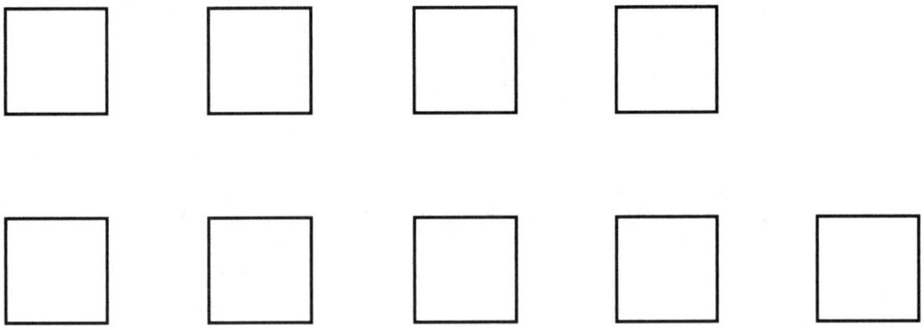

0 1 2 3 4 5 6 7 8 9

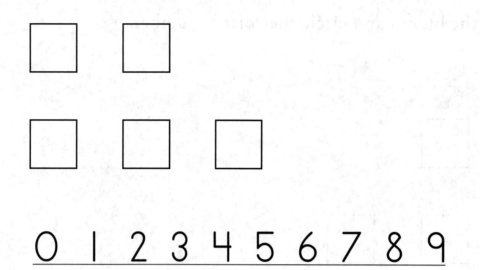

0 1 2 3 4 5 6 7 8 9

SYSTEMATIC REVIEW

Count and circle the correct number.

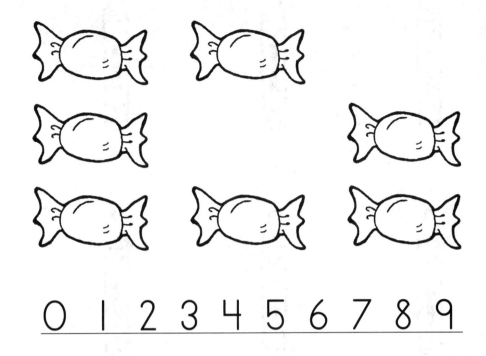

0 1 2 3 4 5 6 7 8 9

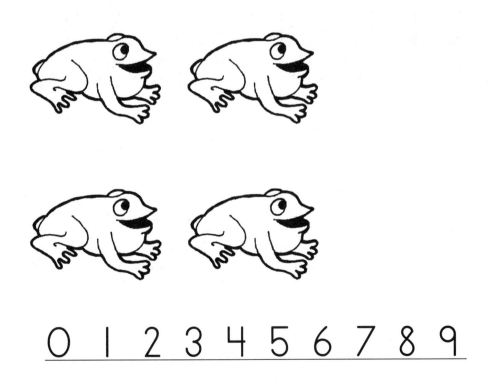

0 1 2 3 4 5 6 7 8 9

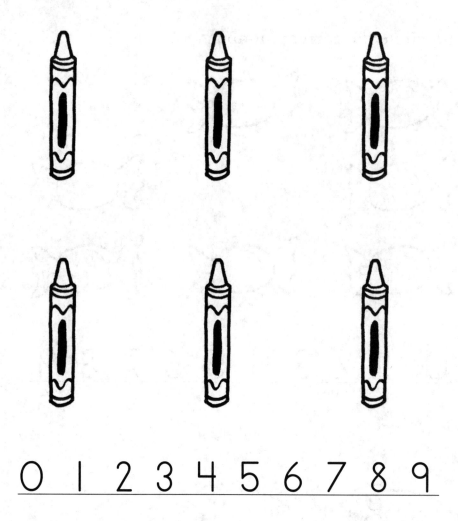

0 1 2 3 4 5 6 7 8 9

SYSTEMATIC REVIEW

Count and circle the correct number.

0 1 2 3 4 5 6 7 8 9

- -

0 1 2 3 4 5 6 7 8 9

0 1 2 3 4 5 6 7 8 9

SYSTEMATIC REVIEW

Count and circle the correct number.

0 1 2 3 4 5 6 7 8 9

0 1 2 3 4 5 6 7 8 9

0 1 2 3 4 5 6 7 8 9

Draw lines to match the numbers with the pictures. The first one has been done for you.

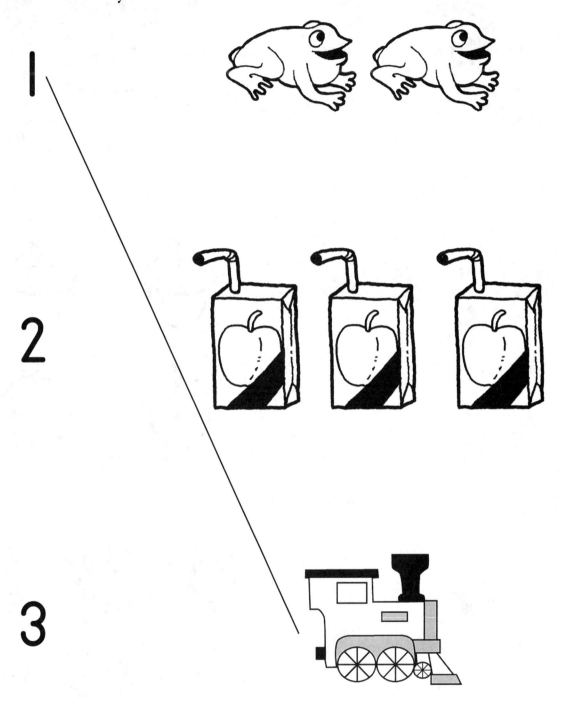

Draw lines to match the numbers with the pictures.

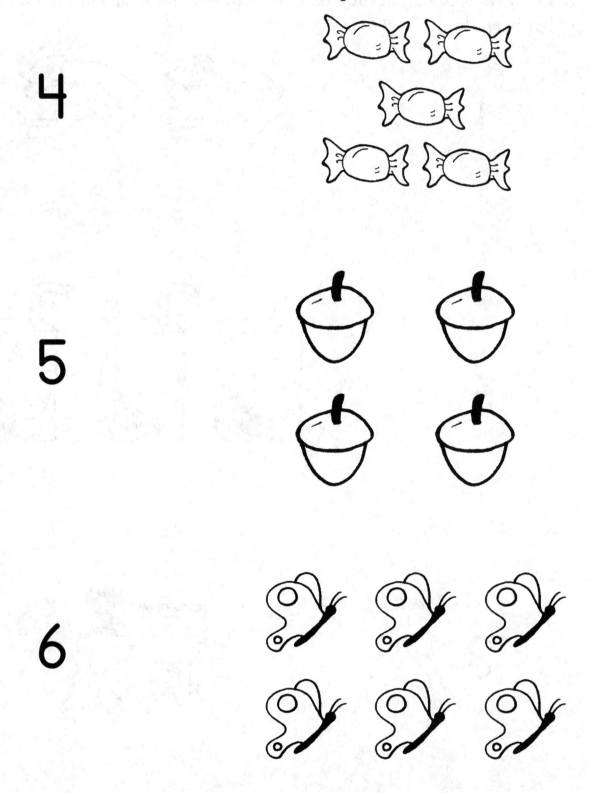

Cross out the wrong answer. Trace the correct answer.

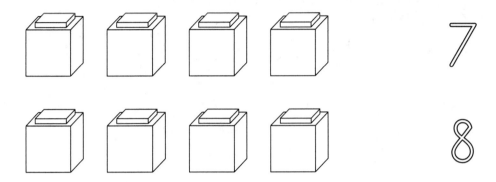

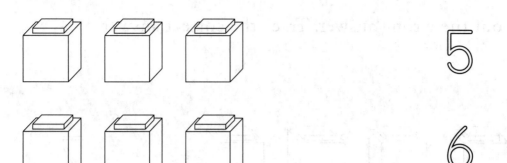

5

6

3

4

2B

Cross out the wrong answer. Trace the correct answer.

2

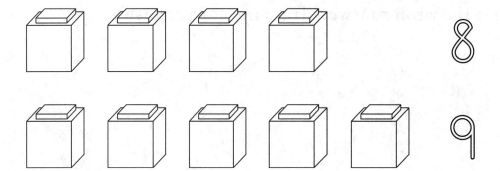

8

9

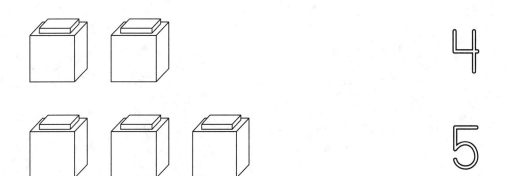

4

5

LESSON PRACTICE

Cross out the wrong answer. Trace the correct answer.

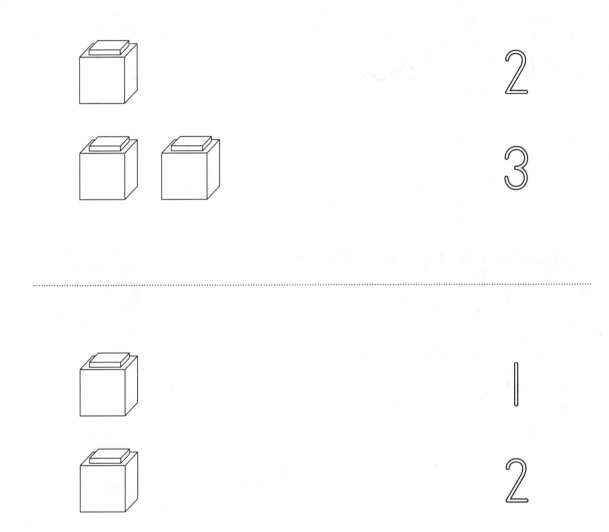

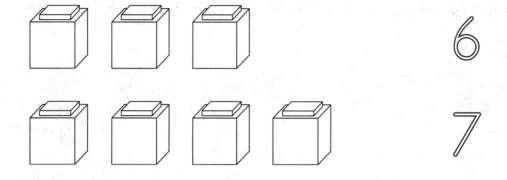

6

7

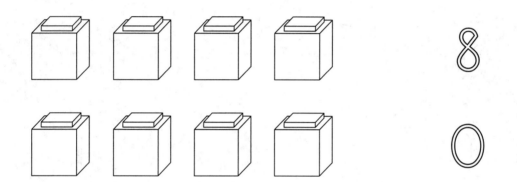

8

0

SYSTEMATIC REVIEW

Cross out the wrong answer. Trace the correct answer.

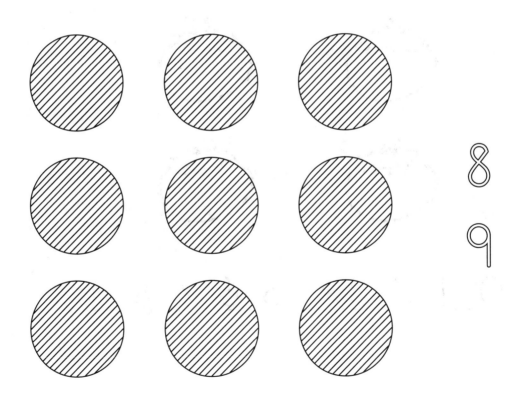

Count and circle the correct number.

0 1 2 3 4 5 6 7 8 9

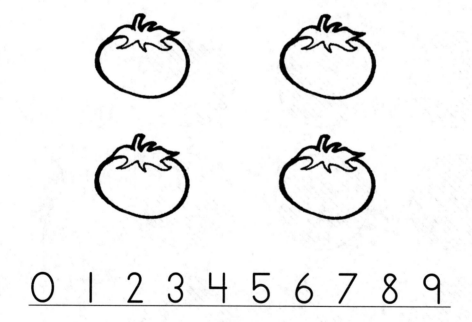

0 1 2 3 4 5 6 7 8 9

SYSTEMATIC REVIEW

Cross out the wrong answer. Trace the correct answer.

1
9

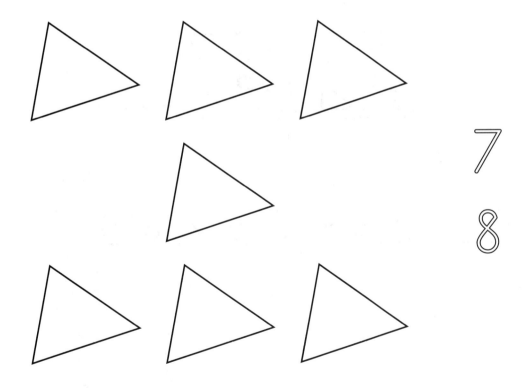

7
8

Count and circle the correct number.

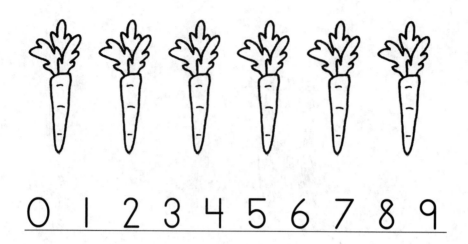

0 1 2 3 4 5 6 7 8 9

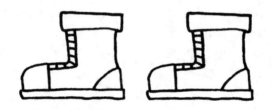

0 1 2 3 4 5 6 7 8 9

SYSTEMATIC REVIEW

Cross out the wrong answer. Trace the correct answer.

Count and circle the correct number.

0 1 2 3 4 5 6 7 8 9

0 1 2 3 4 5 6 7 8 9

Trace each number with your finger. Follow the arrows. Draw a line for each car to drive on.

Trace each number with your finger. Follow the arrows. Draw a line for each truck to drive on.

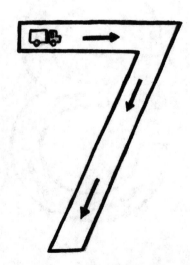

Shade or color the squares to show the correct number. Use your pencil to trace the numerals.

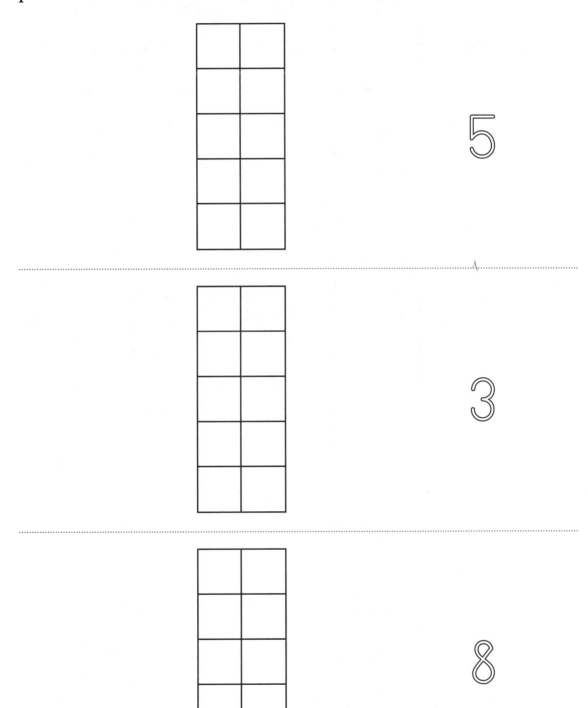

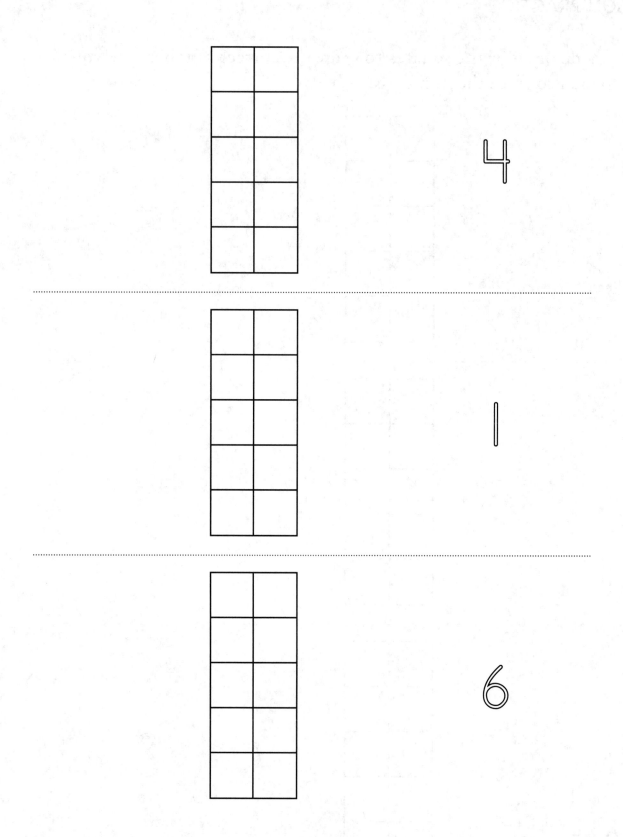

Shade or color the squares to show the correct number. Use your pencil to trace the numerals.

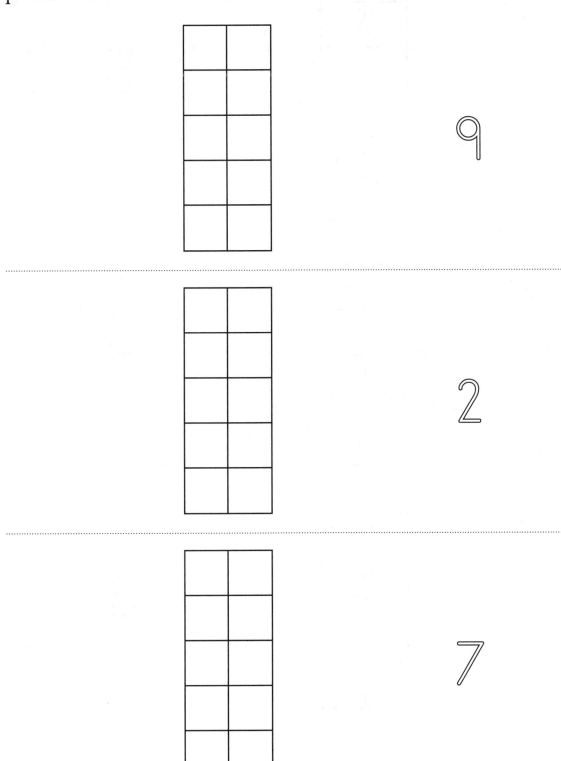

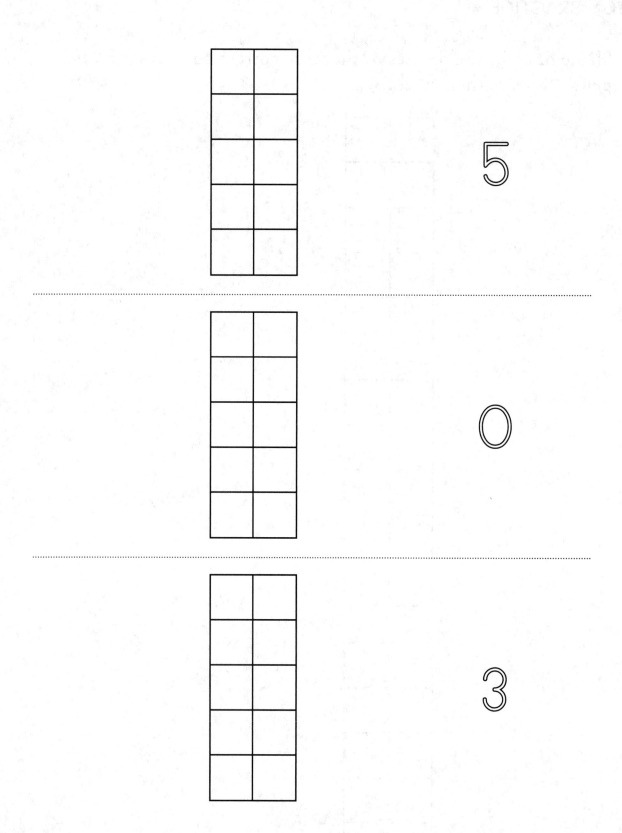

5

0

3

Shade or color the squares to show the correct number. Use your pencil to trace the numerals.

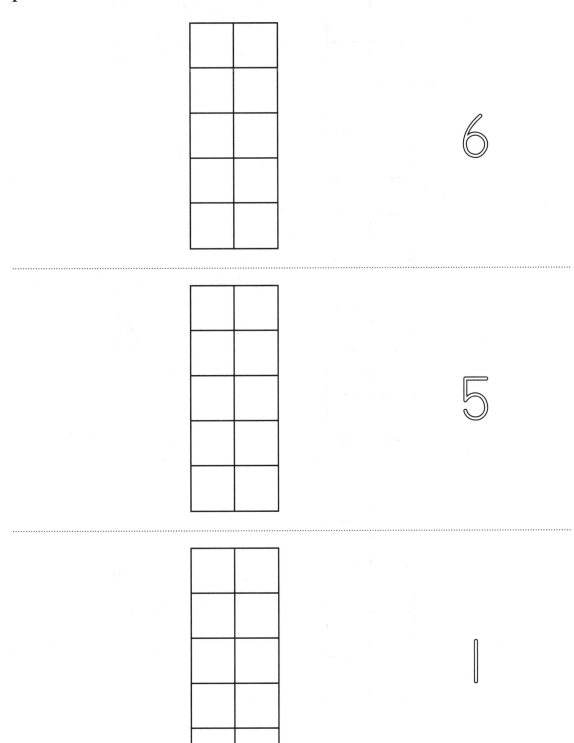

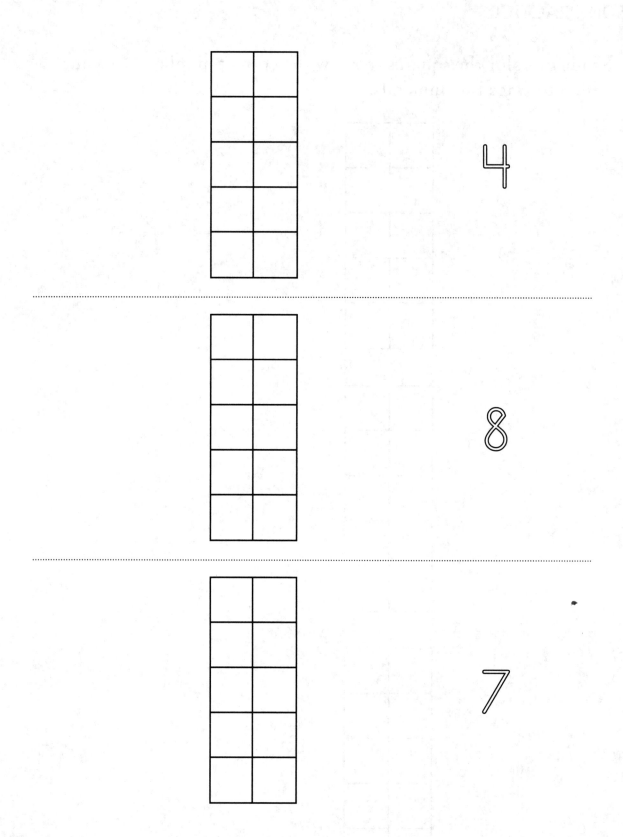

3D

Shade or color the squares to show the correct number. Use your pencil to trace the numerals.

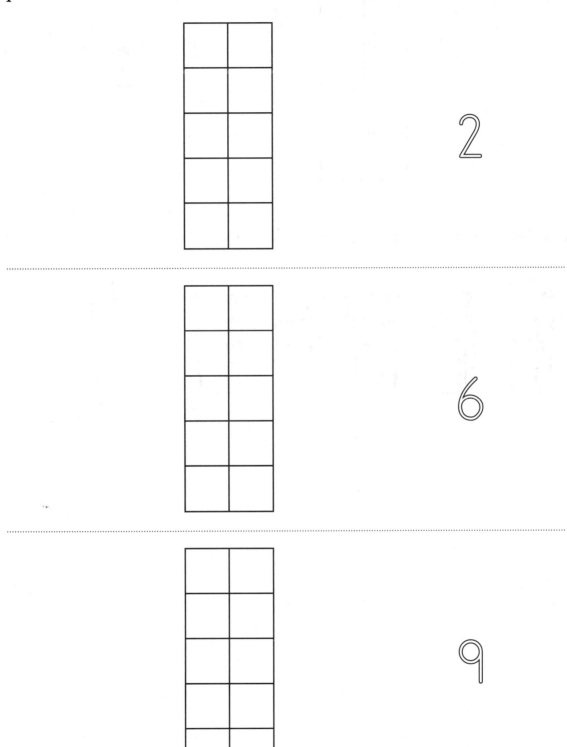

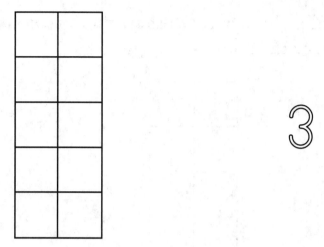

Cross out the wrong answer. Trace the correct answer.

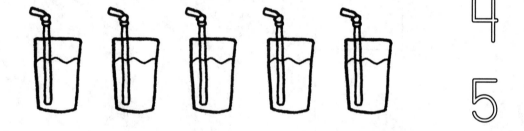

4
5

SYSTEMATIC REVIEW

3E

Shade or color the squares to show the correct number. Use your pencil to trace the numerals.

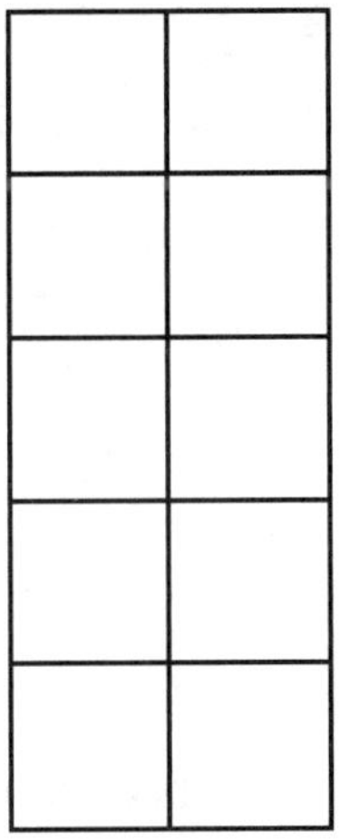

1

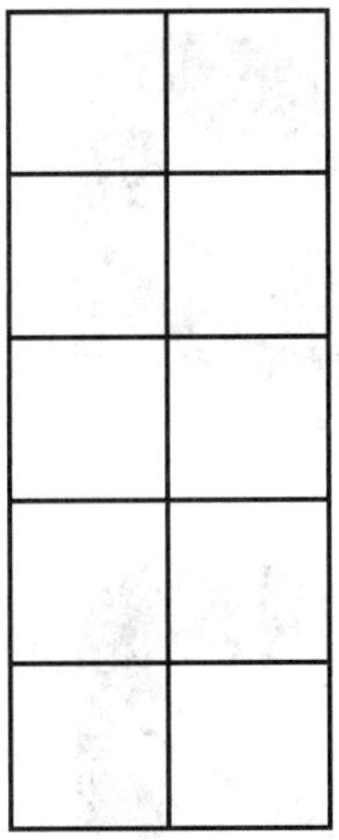

8

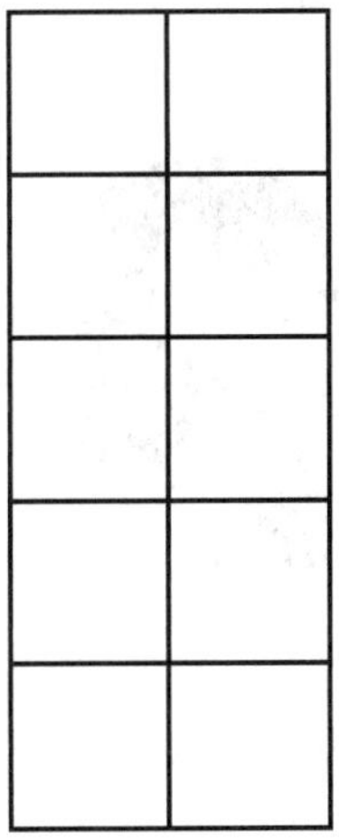

4

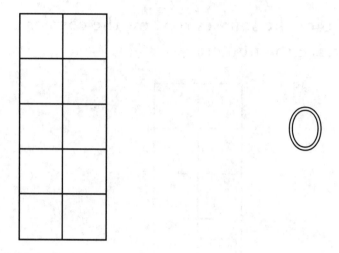

Cross out the wrong answer. Trace the correct answer.

SYSTEMATIC REVIEW

Shade or color the squares to show the correct number. Use your pencil to trace the numerals.

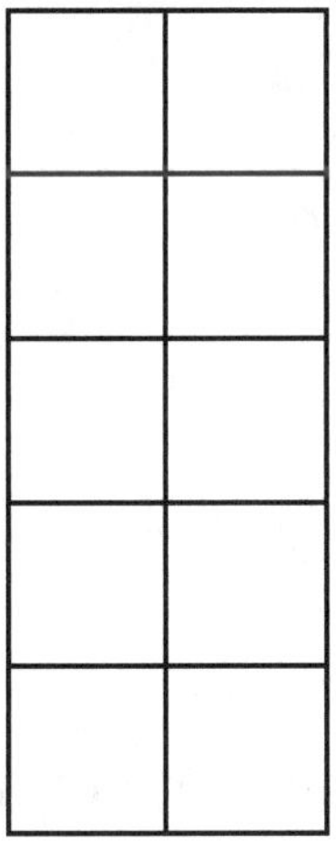

3

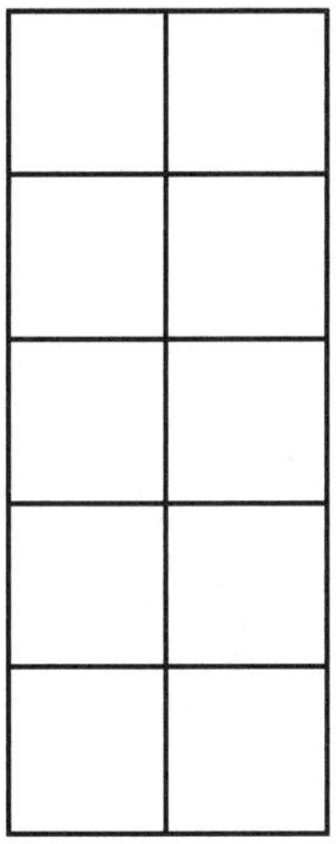

9

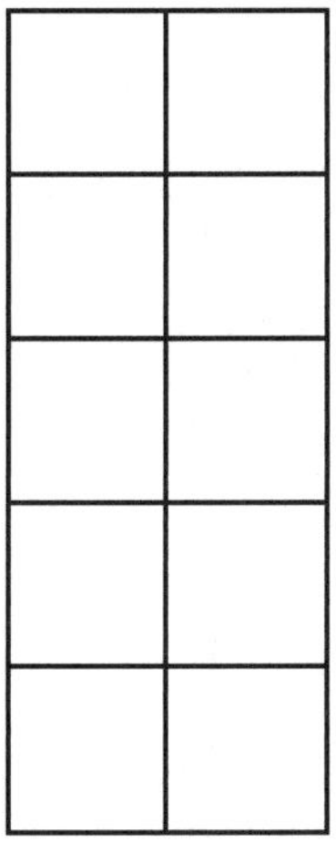

7

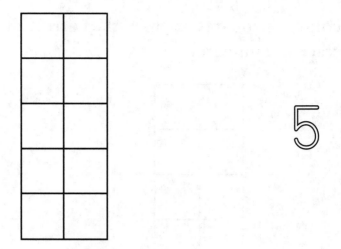

Cross out the wrong answer. Trace the correct answer.

2

3

Draw lines to match the numbers with the pictures.

7

8

9

Trace each number with your finger. Follow the arrows. Draw a line for each bus to drive on.

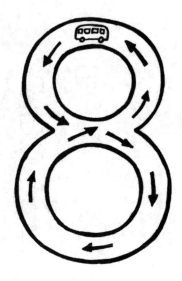

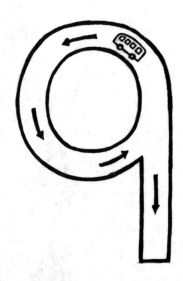

Count the rectangles that match the picture. Circle and say the correct number.

How many rectangles are white? ▭

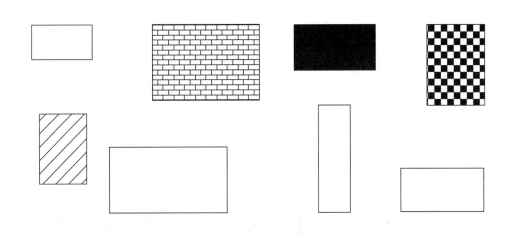

0 1 2 3 4 5 6 7 8 9

How many rectangles are black? ▬

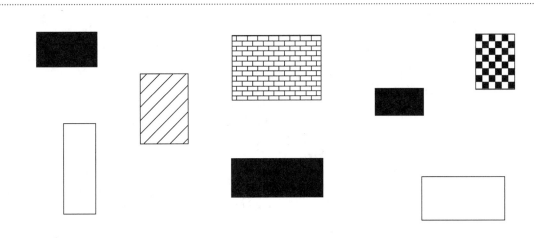

0 1 2 3 4 5 6 7 8 9

How many rectangles have stripes?

0 1 2 3 4 5 6 7 8 9

Count the rectangles that match the picture. Circle and say the correct number.

How many rectangles have bricks?

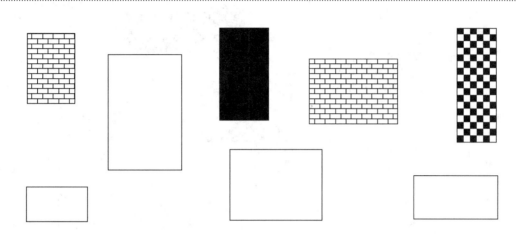

0 1 2 3 4 5 6 7 8 9

How many rectangles are gray?

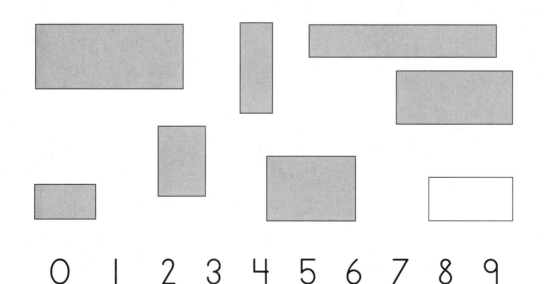

0 1 2 3 4 5 6 7 8 9

How many checkerboard rectangles are there?

0 1 2 3 4 5 6 7 8 9

Count the rectangles that match the picture. Circle and say the correct number.

How many rectangles have spots?

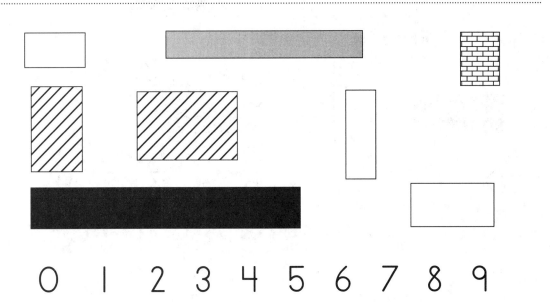

How many rectangles are white?

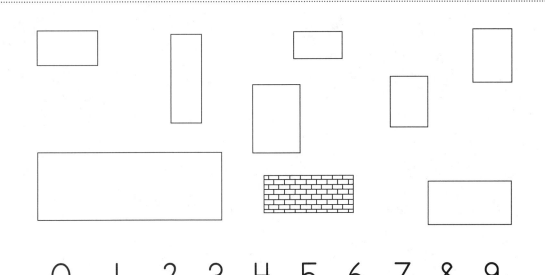

How many rectangles have waves?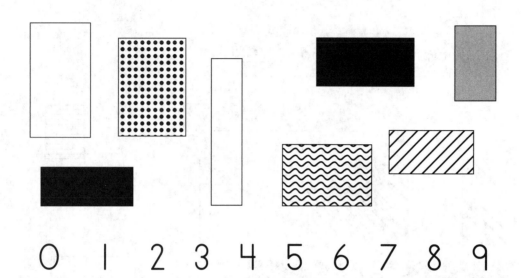

0 1 2 3 4 5 6 7 8 9

Count the rectangles that match the picture. Circle and say the correct number.

How many rectangles have bricks?

0 1 2 3 4 5 6 7 8 9

Cross out the wrong answer. Trace the correct answer.

7

8

Shade or color the squares to show the correct number. Use your pencil to trace the numerals.

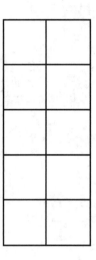

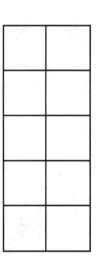

4E

SYSTEMATIC REVIEW

Count the rectangles that match the picture. Circle and say the correct number.

How many rectangles are gray? 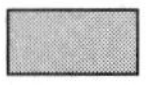

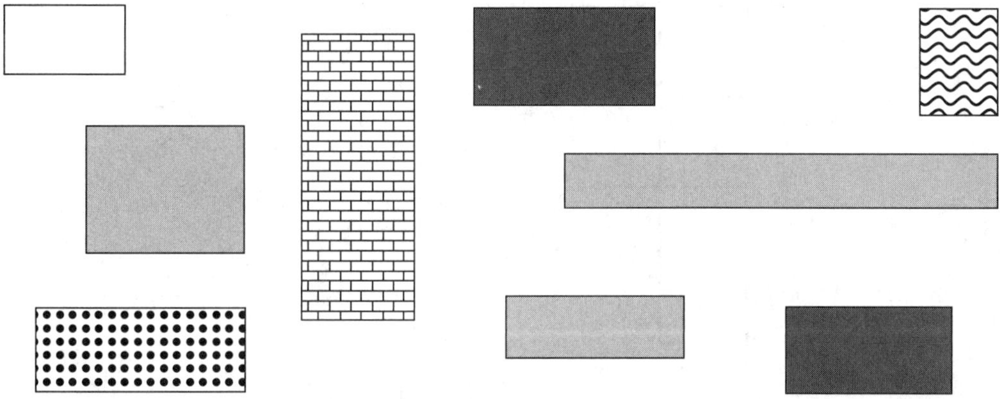

0 1 2 3 4 5 6 7 8 9

Cross out the wrong answer. Trace the correct answer.

0

1

Shade or color the squares to show the correct number. Use your pencil to trace the numerals.

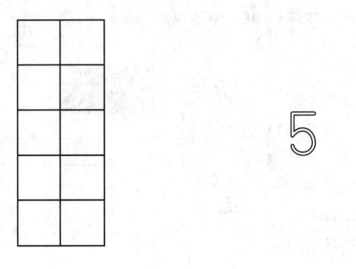

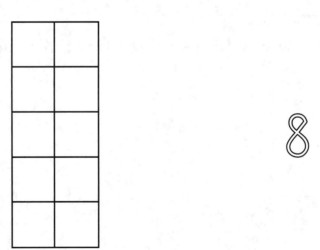

Count the rectangles that match the picture. Circle and say the correct number.

How many rectangles have stripes?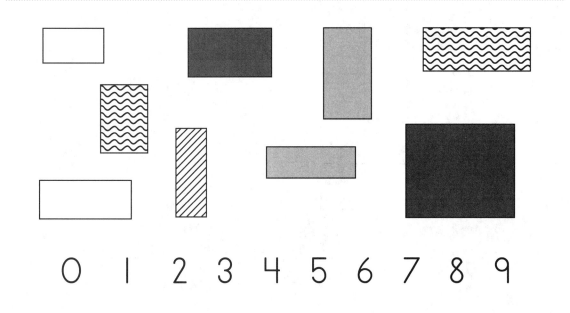

Cross out the wrong answer. Trace the correct answer.

Shade or color the squares to show the correct number. Use your pencil to trace the numerals.

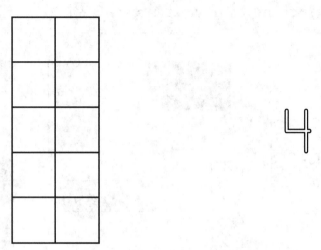

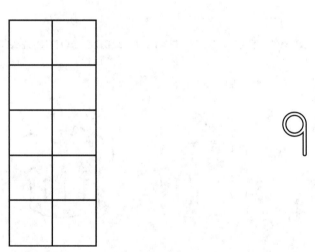

4G

Draw the shape that comes next. Color the shapes if you wish.

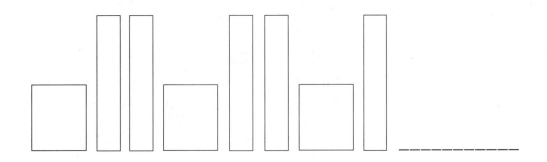

How many rectangles do you see in the house? Circle the answer.

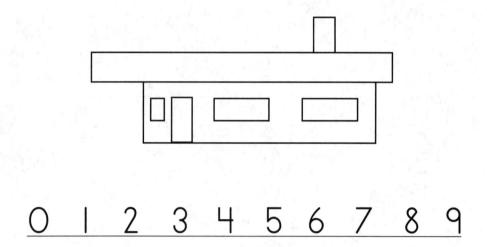

0 1 2 3 4 5 6 7 8 9

How many rectangles do you see in the truck? Circle the answer.

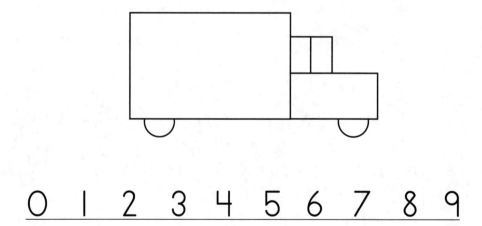

0 1 2 3 4 5 6 7 8 9

Put the blocks on the squares. Count and say the correct number.
Write it if you can.

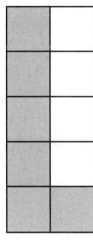

_____

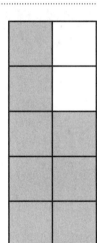

_____

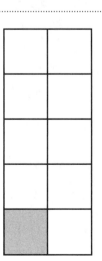

_____

Put the blocks on the squares. Count and say the correct number. Write it if you can.

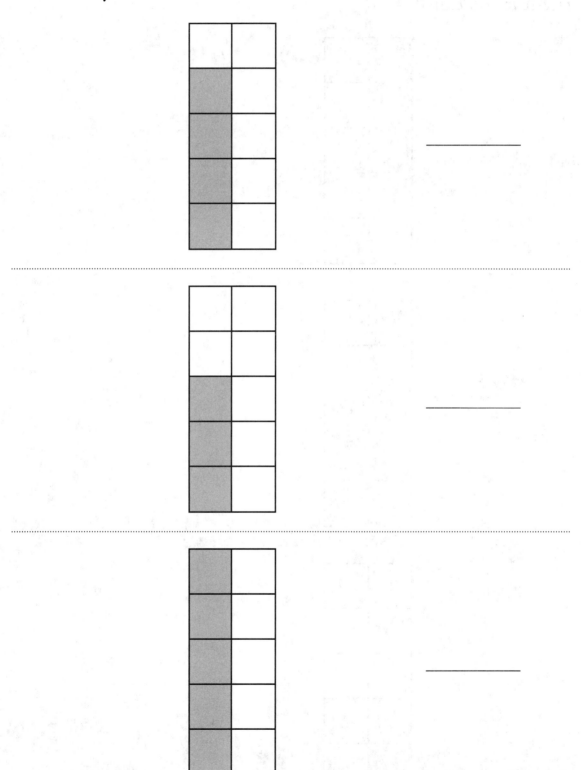

_____

_____

_____

5B

Put the blocks on the squares. Count and say the correct number.
Write it if you can.

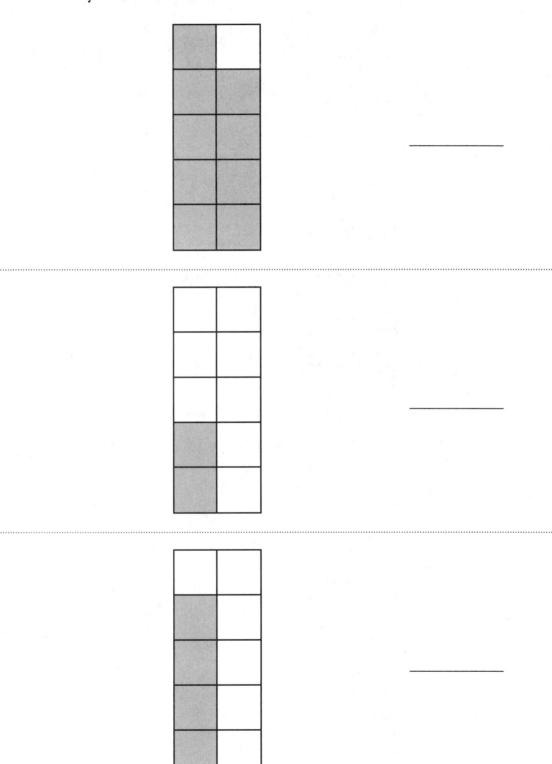

Put the blocks on the squares. Count and say the correct number.
Write it if you can.

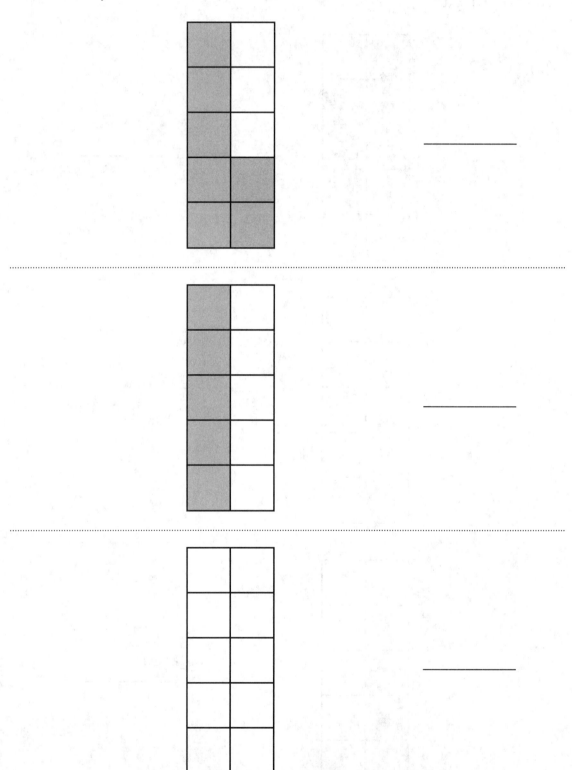

_____

_____

_____

locks on the squares. Count and say the correct number.
f you can.

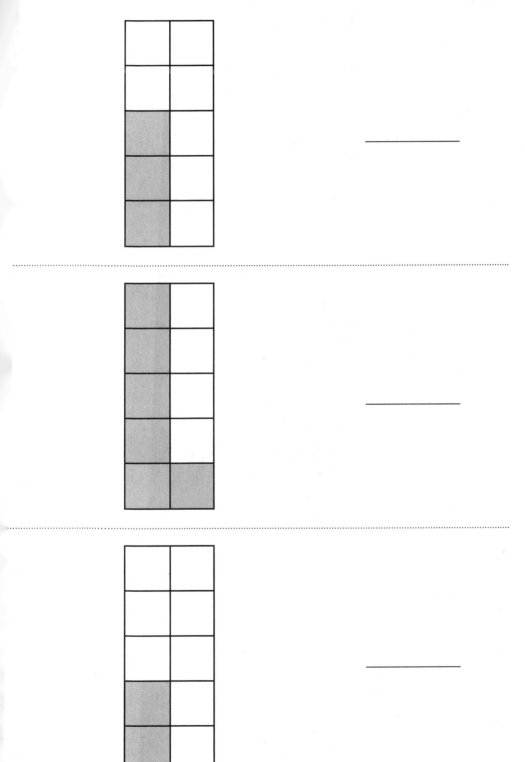

_____

_____

_____

LESSON PRACTICE

Put the blocks on the squares. Count and
Write it if you can.

Put the b
Write it

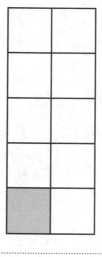

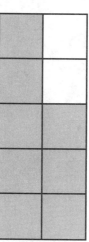

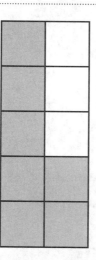

5D

Put the blocks on the squares. Count and say the correct number.
Write it if you can.

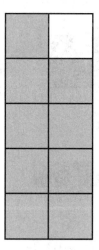

_____

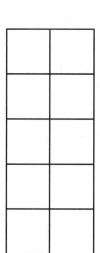

_____

Count the rectangles that match the picture. Circle and say the correct number.

How many rectangles are black? ▬

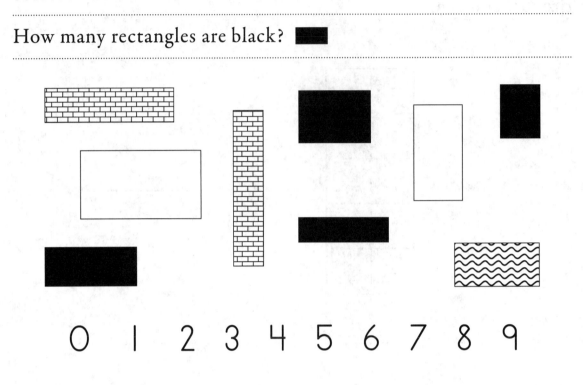

0 1 2 3 4 5 6 7 8 9

Cross out the wrong answer. Trace the correct answer.

Count the rectangles that match the picture. Circle and say the correct number.

How many rectangles are black? ▬

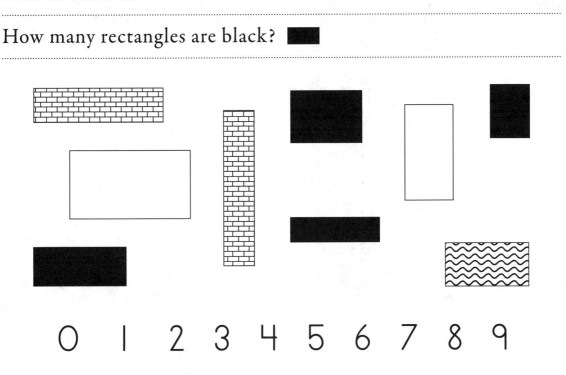

0 1 2 3 4 5 6 7 8 9

Cross out the wrong answer. Trace the correct answer.

SYSTEMATIC REVIEW

Put the blocks on the squares. Count and say the correct number.
Write it if you can.

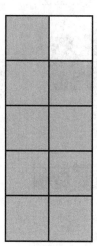

_____

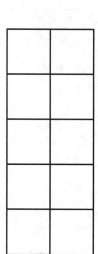

_____

Put the blocks on the squares. Count and say the correct number.
Write it if you can.

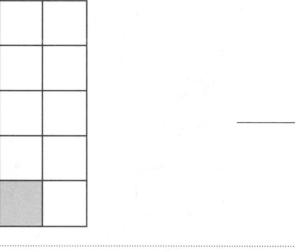

_____

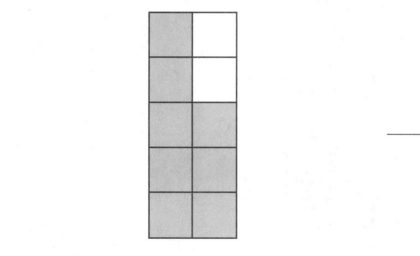

_____

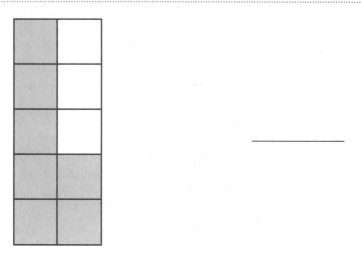

_____

Put the blocks on the squares. Count and say the correct number.
Write it if you can.

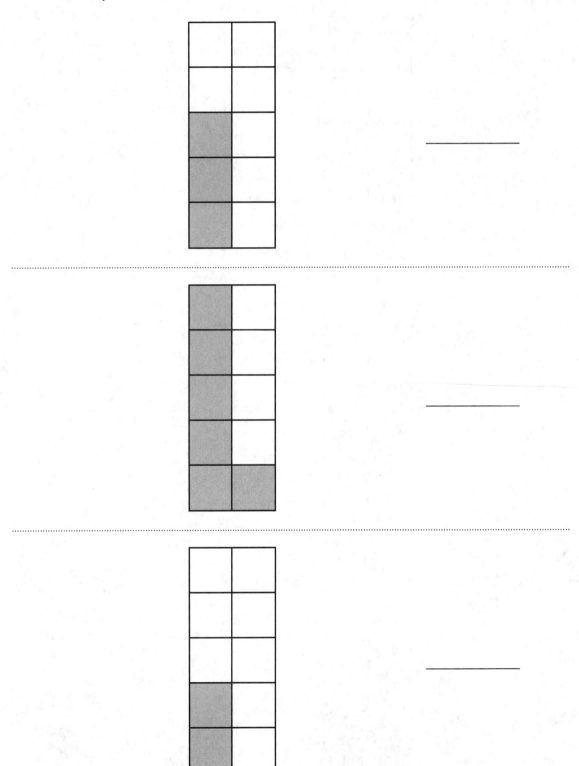

_____

_____

_____

Put the blocks on the squares. Count and say the correct number.
Write it if you can.

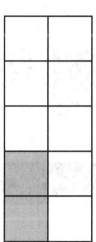

_____

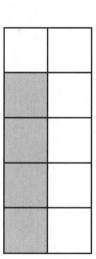

_____

Count the rectangles that match the picture. Circle and say the correct number.

How many checkerboard rectangles are there?

0 1 2 3 4 5 6 7 8 9

Cross out the wrong answer. Trace the correct answer.

O

I

Put the blocks on the squares. Count and say the correct number.
Write it if you can.

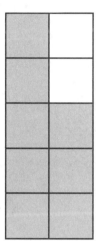

_____

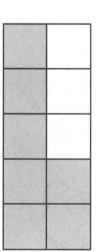

_____

Count the rectangles that match the picture. Circle and say the correct number.

How many rectangles have waves?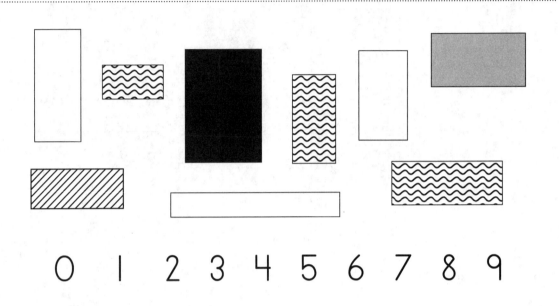

Cross out the wrong answer. Trace the correct answer.

Draw lines to match the numbers with the pictures.

2

4

1

Trace each number with your finger. Follow the arrows. Start at number 1 and draw a line. Then make the bird fly to number 2 and finish the drawing.

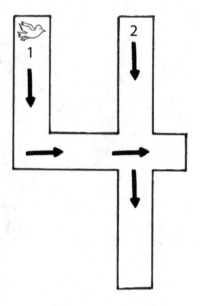

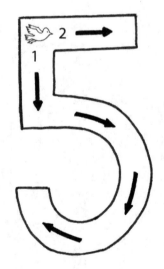

LESSON PRACTICE

Count the circles that match the picture. Circle and say the correct number.

How many circles are white? ○

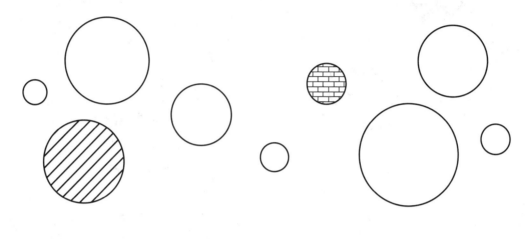

0 1 2 3 4 5 6 7 8 9

How many circles are black?

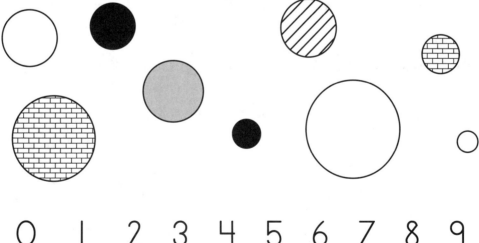

0 1 2 3 4 5 6 7 8 9

How many circles have stripes?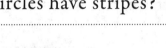

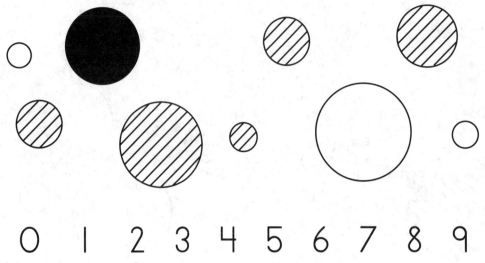

0 1 2 3 4 5 6 7 8 9

Circle the correct answers. Use the blocks to check your answers.

Which number is greater?

6   9

Which number is less?

9   5

LESSON PRACTICE

Count the circles that match the picture. Circle and say the correct number.

How many circles are gray?

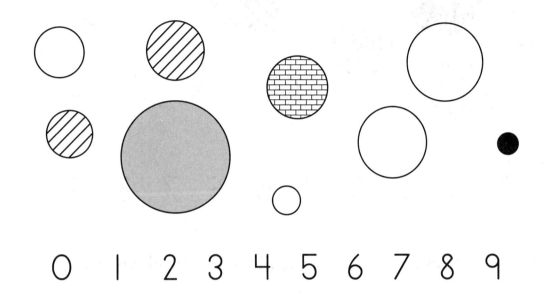

0   1   2   3   4   5   6   7   8   9

How many circles have rectangles in them?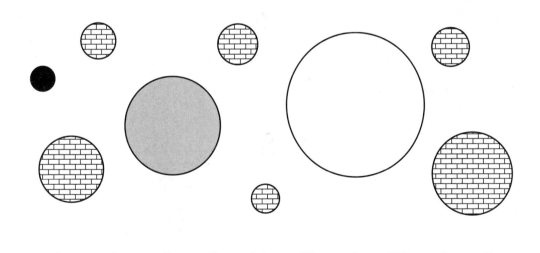

0   1   2   3   4   5   6   7   8   9

How many checkerboard circles are there?

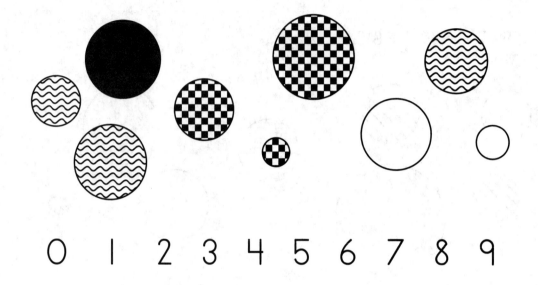

0 1 2 3 4 5 6 7 8 9

Circle the correct answers. Use the blocks to check your answers.

Which number is less?

1   3

Which number is greater?

6   8

Count the circles that match the picture. Circle and say the correct number.

How many circles have waves?

0 1 2 3 4 5 6 7 8 9

How many circles are white? ◯

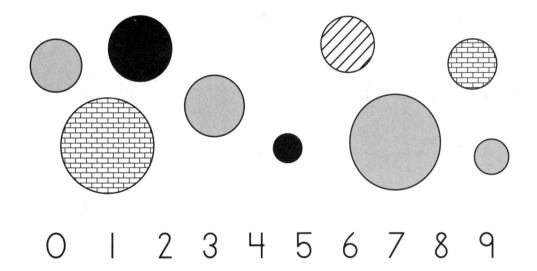

0 1 2 3 4 5 6 7 8 9

How many circles have spots?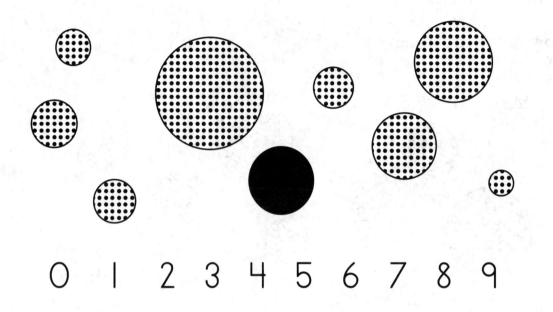

0 1 2 3 4 5 6 7 8 9

Circle the correct answers. Use the blocks to check your answers.

Which number is greater?

0   6

Which number is less?

5   4

Count the circles that match the picture. Circle and say the correct number.

How many circles have stripes?

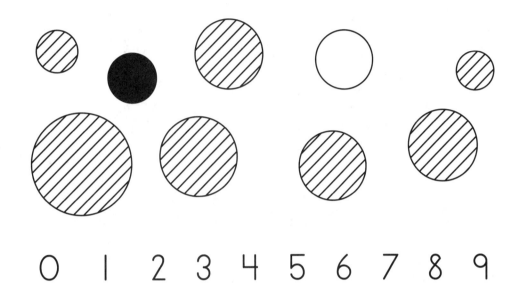

0 1 2 3 4 5 6 7 8 9

How many rectangles are black? ▬

0 1 2 3 4 5 6 7 8 9

Put the blocks on the squares. Count and say the correct number.
Write it if you can.

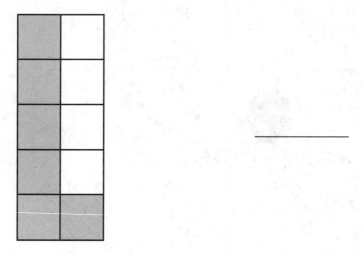

_____

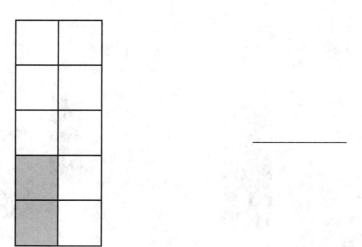

_____

Count the circles that match the picture. Circle and say the correct number.

How many circles have rectangles in them?

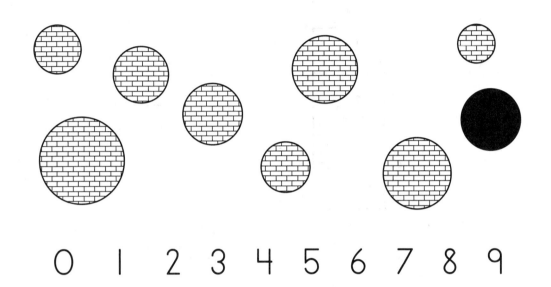

0 1 2 3 4 5 6 7 8 9

How many rectangles are gray? ▭

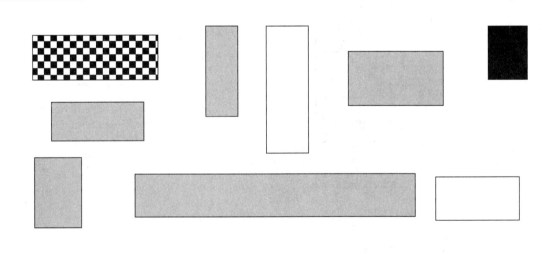

0 1 2 3 4 5 6 7 8 9

Put the blocks on the squares. Count and say the correct number.
Write it if you can.

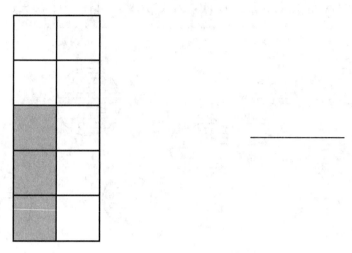

_____

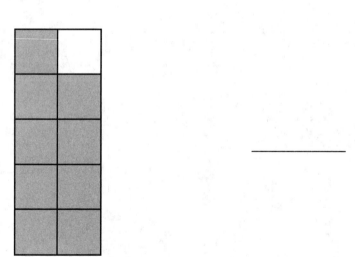

_____

SYSTEMATIC REVIEW

Count the circles that match the picture. Circle and say the correct number.

How many circles have waves?

How many checkerboard rectangles are there?

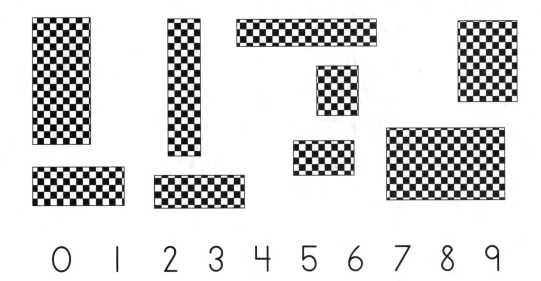

Put the blocks on the squares. Count and say the correct number.
Write it if you can.

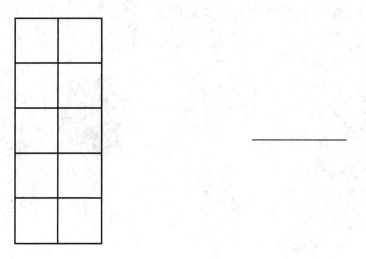

_____

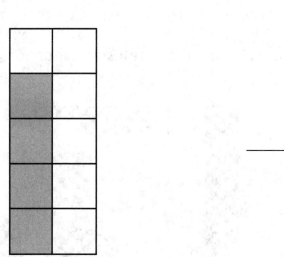

_____

Draw the shape that comes next. Color the shapes if you wish.

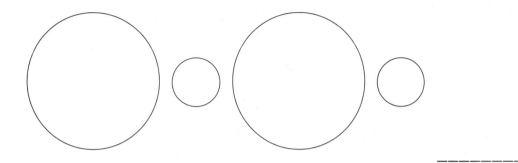

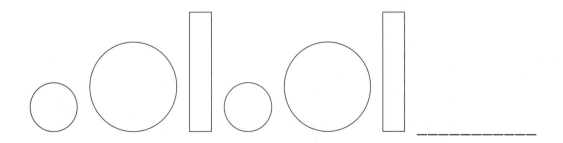

Dad made two piles of blocks. Circle the pile that has *more* blocks.

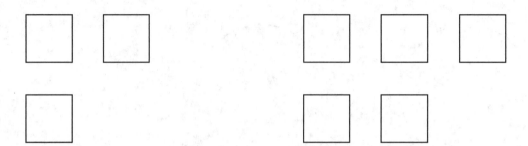

Mom poured two cups of milk. Circle the cup that has *less* milk.

Color the *shorter* block yellow. Color the *longer* block purple.

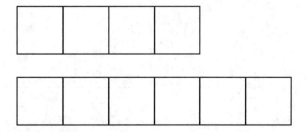

7A

Color the squares or write the number.

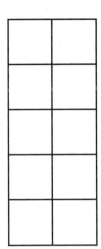

_____

_____

Color the squares or write the number.

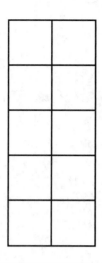

7

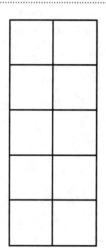

2

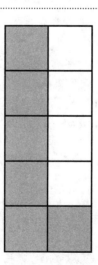

_____

90

PRIMER

Color the squares or write the number.

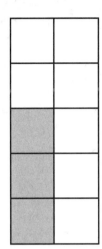

_____

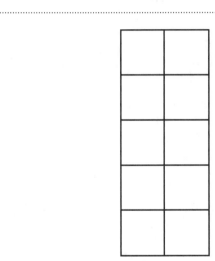

0

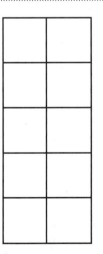

4

Color the squares or write the number.

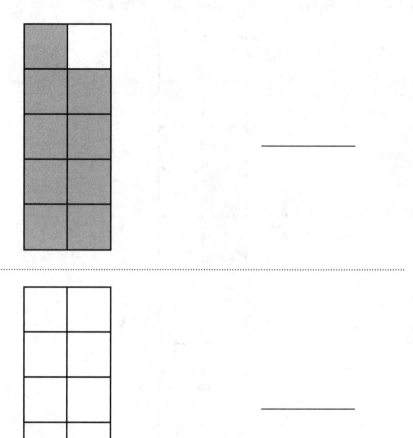

_____

_____

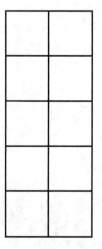

LESSON PRACTICE

Color the squares or write the number.

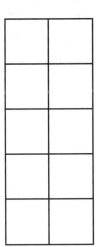

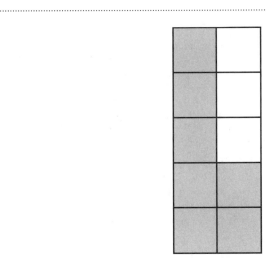

_____

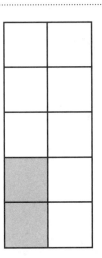

_____

Color the squares or write the number.

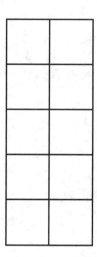

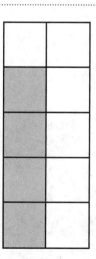

_____

SYSTEMATIC REVIEW

Color the squares or write the number.

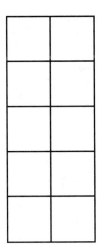

9

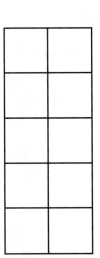

_____

Count the shapes that match the picture. Circle and say the correct number.

How many are rectangles? ▭

0 1 2 3 4 5 6 7 8 9

Cross out the wrong answer. Trace the correct answer.

4

5

SYSTEMATIC REVIEW

Color the squares or write the number.

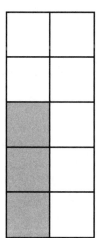

_____

|

Count the shapes that match the picture. Circle and say the correct number.

How many are circles? ○

0 1 2 3 4 5 6 7 8 9

Cross out the wrong answer. Trace the correct answer.

6

7

SYSTEMATIC REVIEW

Color the squares or write the number.

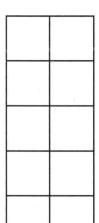

4

_____

Count the shapes that match the picture. Circle and say the correct number.

How many are rectangles? ▭

0 1 2 3 4 5 6 7 8 9

Cross out the wrong answer. Trace the correct answer.

Draw lines to match the numbers with the pictures.
Color the pictures if you wish.

3

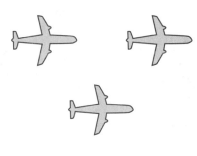

5

8

Draw lines to match the numbers with the pictures.
Color the pictures if you wish.

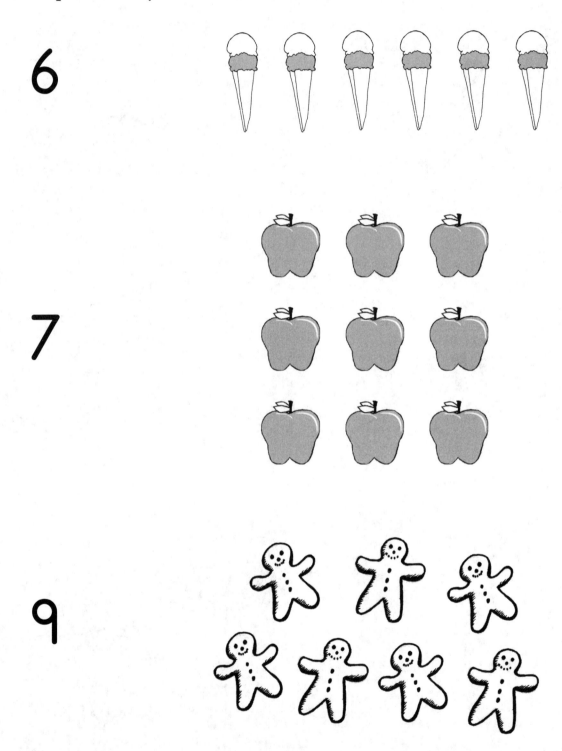

LESSON PRACTICE

Count the triangles that match the picture. Circle and say the correct number.

How many triangles are white? △

0 1 2 3 4 5 6 7 8 9

How many triangles are black? ▲

0 1 2 3 4 5 6 7 8 9

How many triangles have stripes?

0 1 2 3 4 5 6 7 8 9

LESSON PRACTICE

Count the triangles that match the picture. Circle and say the correct number.

How many triangles are gray?

0 1 2 3 4 5 6 7 8 9

How many triangles have waves?

How many triangles have spots?

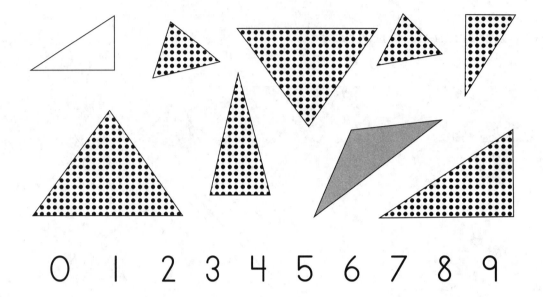

0 1 2 3 4 5 6 7 8 9

Count the triangles that match the picture. Circle and say the correct number.

How many triangles have rectangles in them?

0 1 2 3 4 5 6 7 8 9

How many triangles are white?

0 1 2 3 4 5 6 7 8 9

How many checkerboard triangles are there?

0 1 2 3 4 5 6 7 8 9

Count the triangles that match the picture. Circle and say the correct number.

How many triangles are black?

0 1 2 3 4 5 6 7 8 9

Color the blocks or write the number.

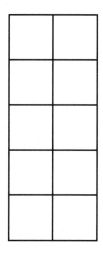

5

Color the blocks or write the number.

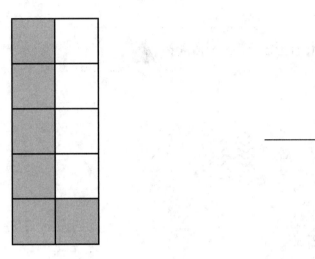

_____

Count the shapes. Circle and say the correct number.

How many are triangles? △

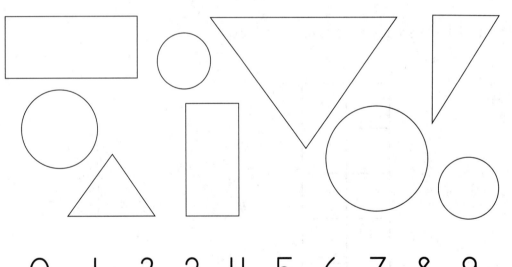

0 1 2 3 4 5 6 7 8 9

8E

Count the triangles that match the picture. Circle and say the correct number.

How many triangles have stripes?

$$0 \quad 1 \quad 2 \quad 3 \quad 4 \quad 5 \quad 6 \quad 7 \quad 8 \quad 9$$

Color the blocks or write the number.

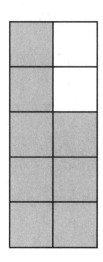

_____

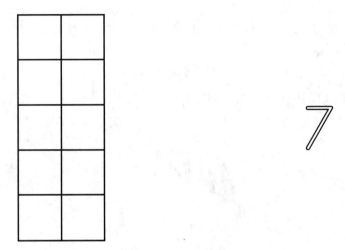

Count the shapes. Circle and say the correct number.

How many are rectangles?

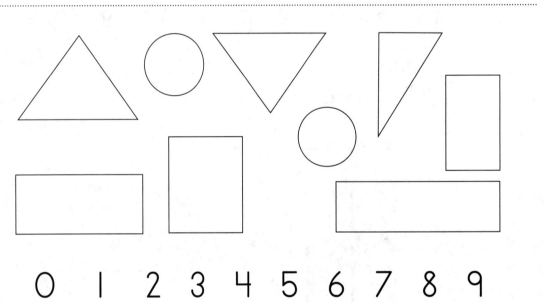

0 1 2 3 4 5 6 7 8 9

Count the triangles that match the picture. Circle and say the correct number.

How many triangles are gray?

0 1 2 3 4 5 6 7 8 9

Color the blocks or write the number.

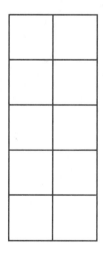

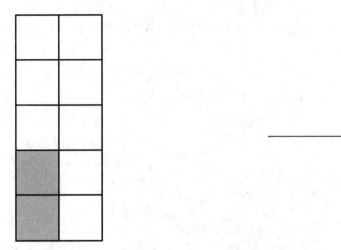

Count the shapes. Circle and say the correct number.

How many are circles? ◯

0 1 2 3 4 5 6 7 8 9

Draw the shape that comes next. Color the shapes if you wish.

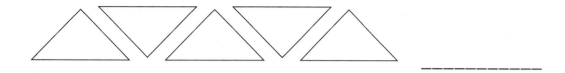 _____

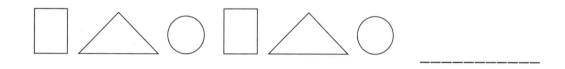 _____

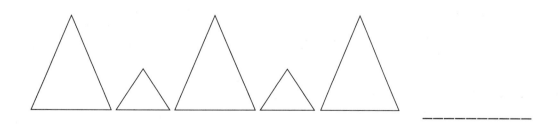 _____

Color the circles blue.
Color the rectangles brown.
Color the little triangles brown.
Color the big triangle red.

Enjoy your sail!

How many circles do you see? _____

How many triangles do you see? _____

How many rectangles do you see? _____

Color the correct number of blocks. Say the number.

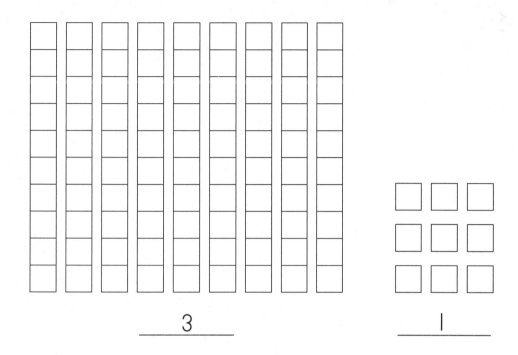

_____3_____   _____1_____

Count and write. Say the number.

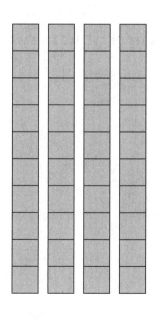

_____   _____

Build and say the numbers.

89

57

Circle the correct answers. Use the blocks to check your answers.

Which number is greater?

3   7

Which number is less?

0   5

9B

Color the correct number of blocks. Say the number.

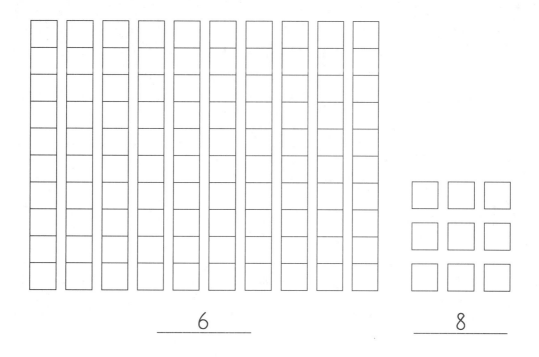

_____6_____   _____8_____

Count and write. Say the number.

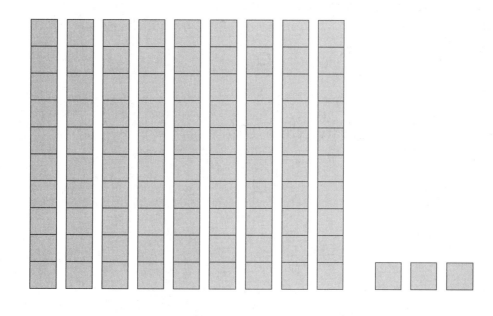

_____   _____

Build and say the numbers.

25

71

Circle the correct answers. Use the blocks to check your answers.

Which number is less?

1   9

Which number is greater than 10?

4   14   5

LESSON PRACTICE

Color the correct number of blocks. Say the number.

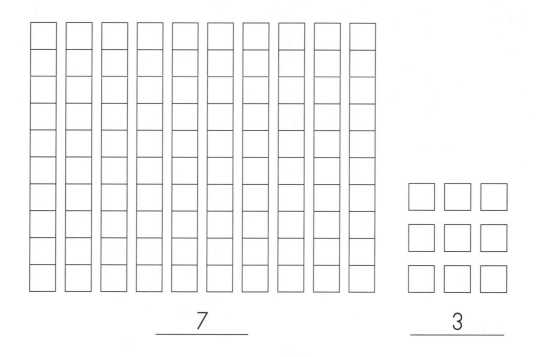

_____7_____   _____3_____

Count and write. Say the number.

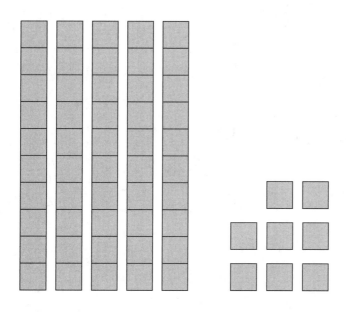

_____   _____

Build and say the numbers.

46

32

- -

Circle the correct answers. Use the blocks to check your answers.

Which number is greater?

7   3

Which numbers are less than 10?

12   8   2

Color the correct number of blocks. Say the number.

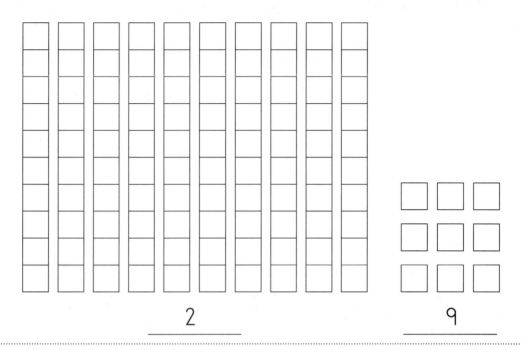

2

9

Build and say the numbers.

81

74

Count the shapes. Circle and say the correct number.

How many are triangles? △

0 1 2 3 4 5 6 7 8 9

Circle the correct answers. Use the blocks to check your answers.

Which number is greater?

4   0

Which number is less?

10   0

Which numbers are less than 10?

9   13   7

Count and write. Then say the number.

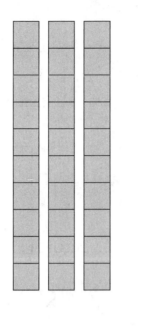

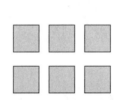

_____   _____

Build and say the numbers.

55

63

Count the shapes. Circle and say the correct number.

How many are rectangles? ▭

0 1 2 3 4 5 6 7 8 9

Circle the correct answers. Use the blocks to check your answers.

Which number is less?

9   2

Which number is greater?

51   15

Which numbers are greater than 10?

3   19   44

Color the correct number of blocks. Say the number.

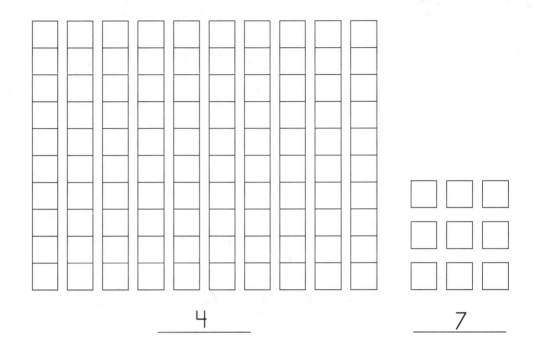

_4_   _7_

....

Build and say the numbers.

92

28

Count the shapes. Circle and say the correct number.

How many are circles? ○

0   1   2   3   4   5   6   7   8   9

Circle the correct answers. Use the blocks to check your answers.

Which number is greater?

5   7

Which number is less?

10   0

Which numbers are less than 10?

0   1   11

Color the numbers green for the units house.

These number cards may be used to play the game described in lesson 9 of the *Primer Instruction Manual.* You may want to laminate the pages before you cut out the cards.

0

1 2 3

4 5 6

7 8 9

Each number pattern corresponds to
the number on the front of the card.
Match the dots with other number cards
or with the blocks to practice number
recognition and counting.

Color the numbers blue for the tens house.

These number cards and the ones on the previous pages may be used to play the game described in lesson 9 of the *Primer Instruction Manual.* You may want to laminate the pages before you cut out the cards.

0		
1	2	3
4	5	6
7	8	9

Each number pattern corresponds to
the number on the front of the card.
Match the dots with other number cards
or with the blocks to practice number
recognition and counting.

Color the correct number of blocks. Say the number.

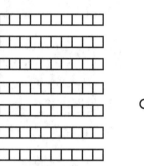

—

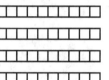

3

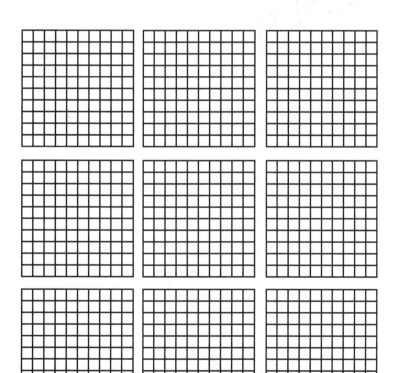

4

Count and write. Say the number.

_____   _____   _____

LESSON PRACTICE

Color the correct number of blocks. Say the number.

8

7

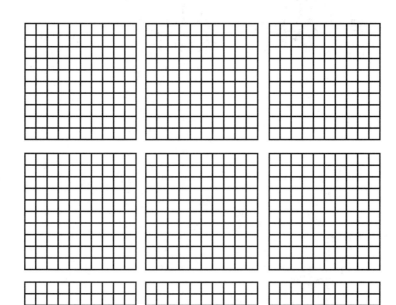

Count and write. Say the number.

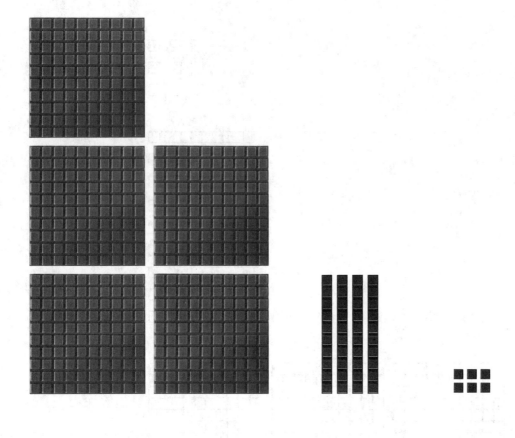

_____   _____   _____

LESSON PRACTICE

Color the correct number of blocks. Say the number.

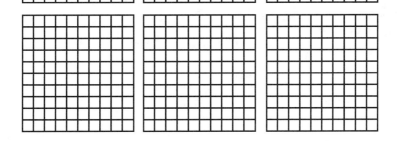

8

3

2

Count and write. Say the number.

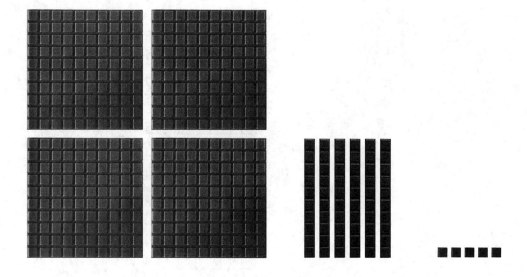

_____   _____   _____

SYSTEMATIC REVIEW

Color the correct number of blocks. Say the number.

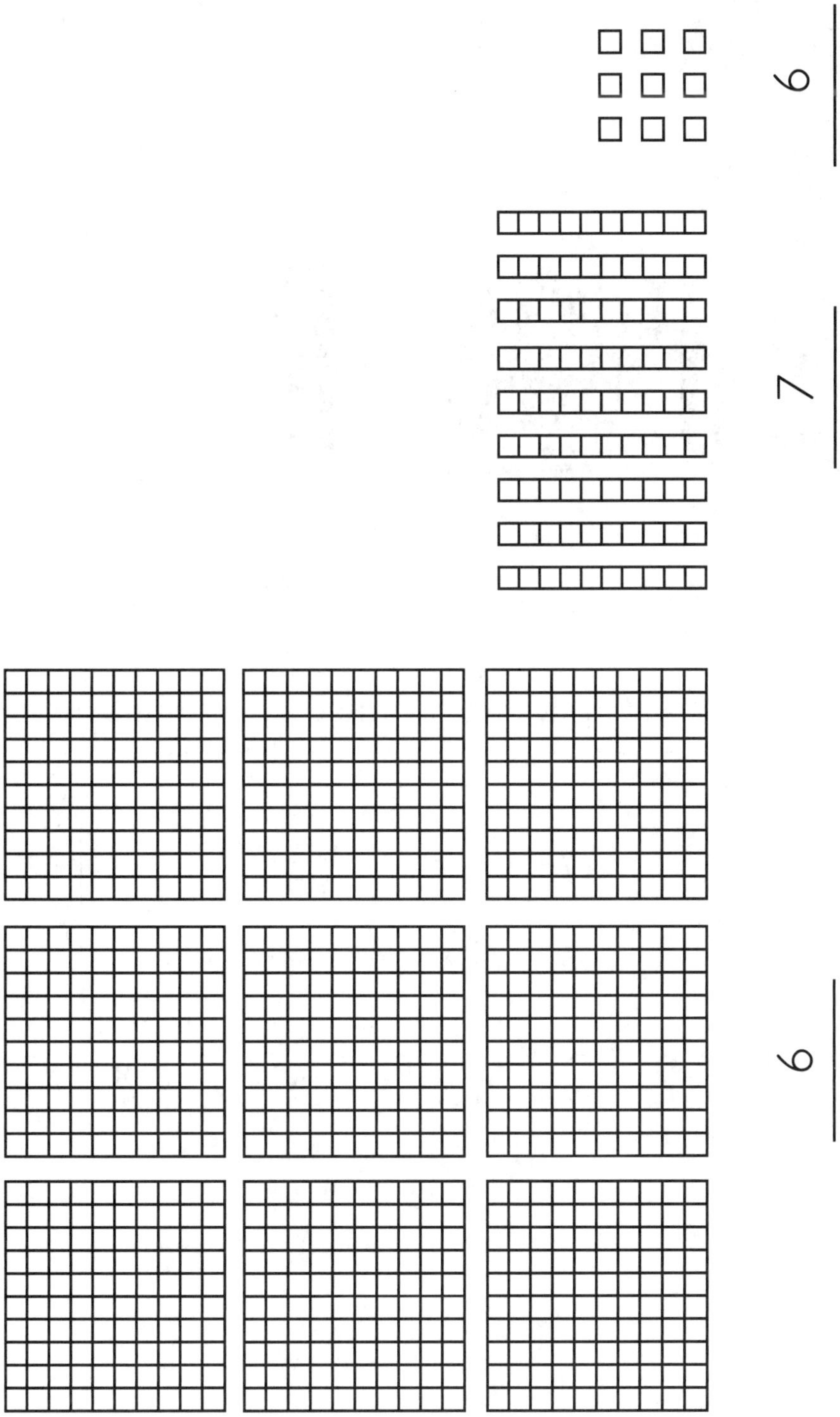

6

7

6

PRIMER SYSTEMATIC REVIEW 10D 139

Count and write. Say the number.

_____   _____   _____

Count the rectangles. Circle and say the correct number. ▭

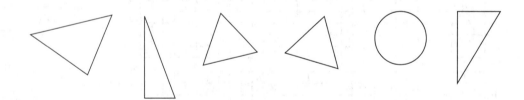

0 1 2 3 4 5 6 7 8 9

SYSTEMATIC REVIEW

Color the correct number of blocks. Say the number.

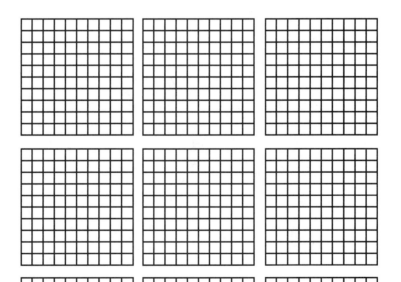

2

3

7

Count and write. Say the number.

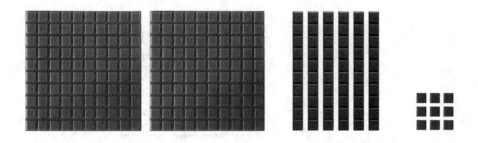

_____   _____ _____

Count the circles. Circle and say the correct number. ○

0 1 2 3 4 5 6 7 8 9

SYSTEMATIC REVIEW

Color the correct number of blocks. Say the number.

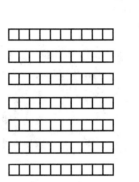

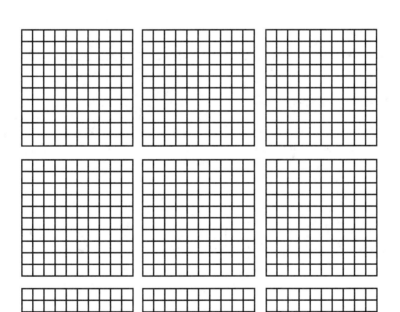

Count and write. Say the number.

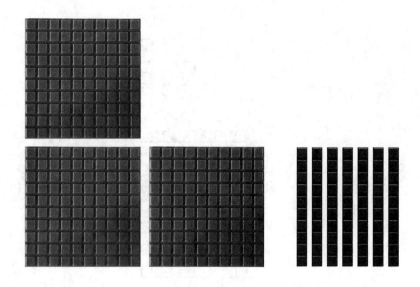

_____   _____   _____

Count the triangles. Circle and say the correct number. ▷

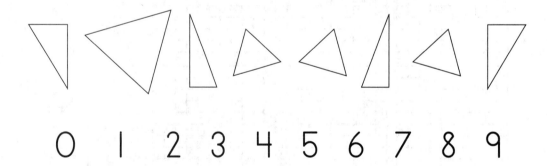

0 1 2 3 4 5 6 7 8 9

Color the numbers red for the hundreds house.

These number cards may be used with the units and tens cards to play the game described in lesson 10 of the *Primer Instruction Manual.* You may want to laminate the pages before you cut out the cards.

0

1	2	3
4	5	6
7	8	9

Each number pattern corresponds to
the number on the front of the card.
Match the dots with other number cards
or with the blocks to practice number
recognition and counting.

Put the unit bars on the pictures. Count, write, and say the numbers. Color the pictures to match the unit bars.

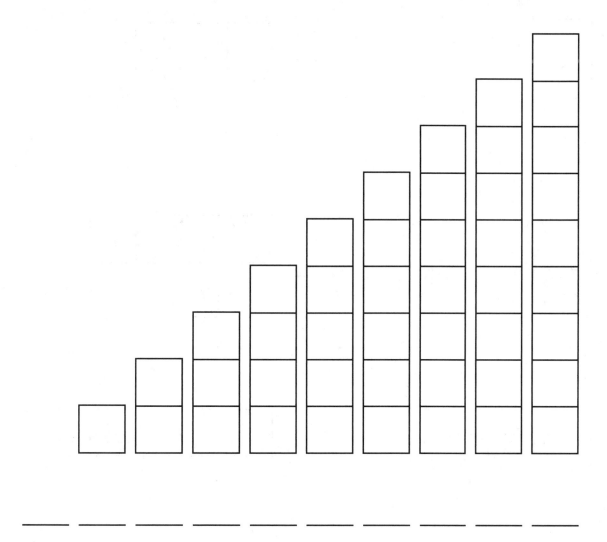

— — — — — — — — —

Count, match, and color. Write the number and say it.

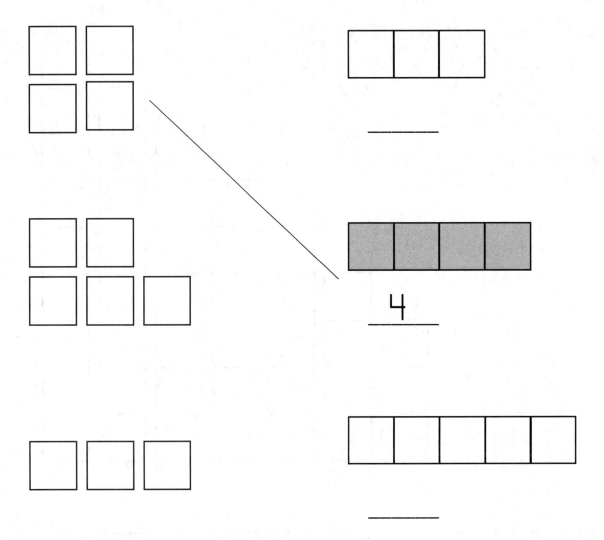

LESSON PRACTICE

Count, match, and color. Write the number and say it.

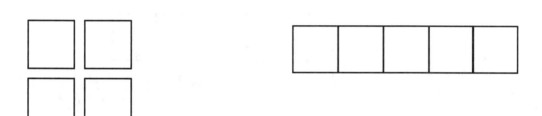

_____

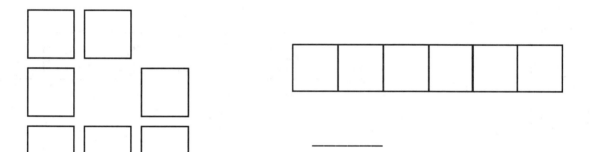

_____

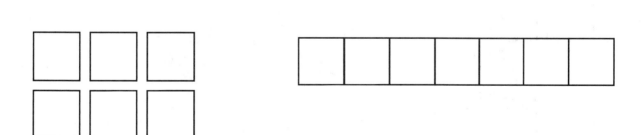

_____

_____

Count, match, and color. Write the number and say it.

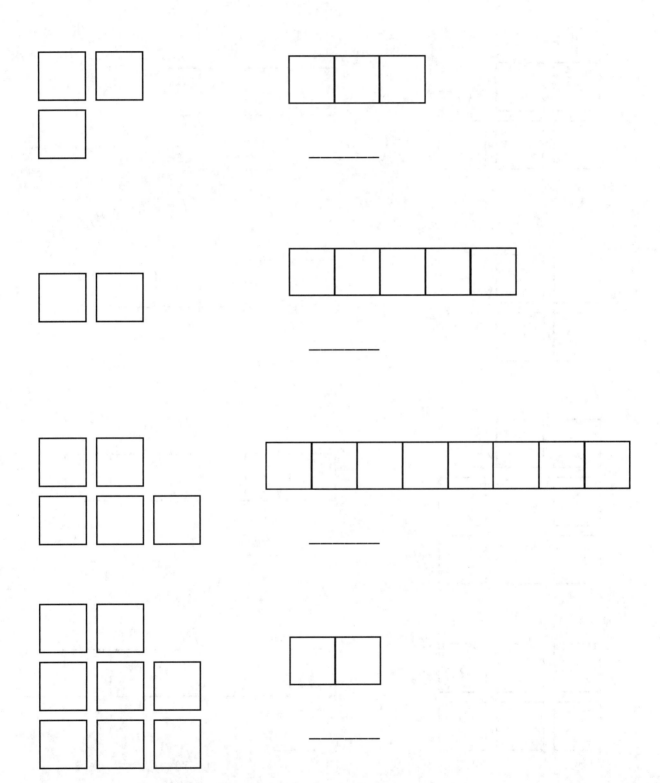

Count, match, and color. Write the number and say it.

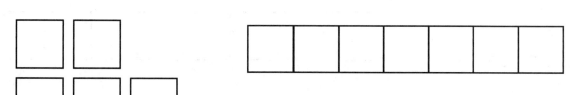

——————

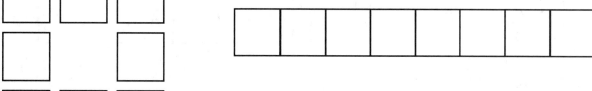

——————

——————

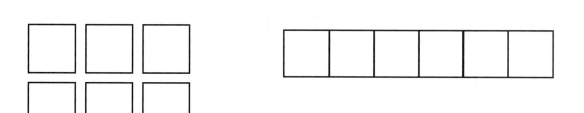

——————

Count, match, and color. Write the number and say it.

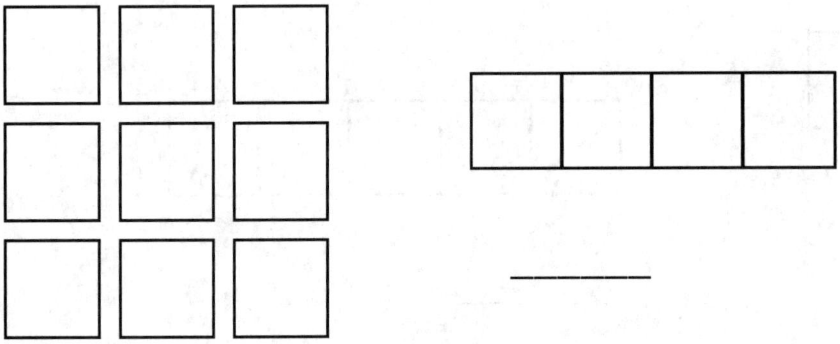

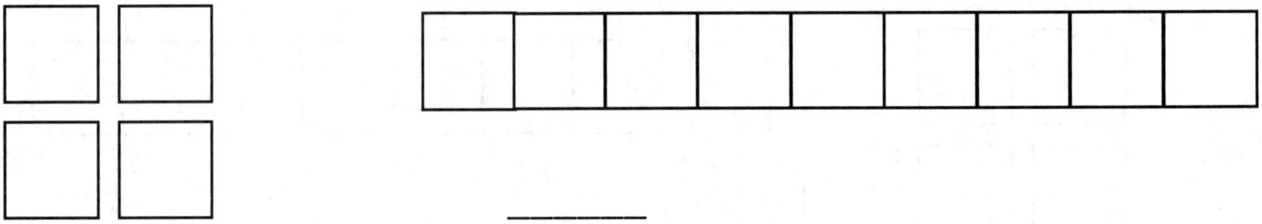

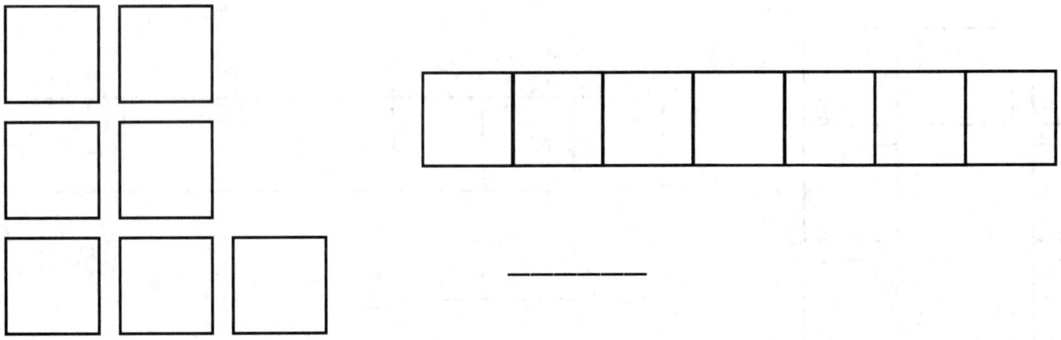

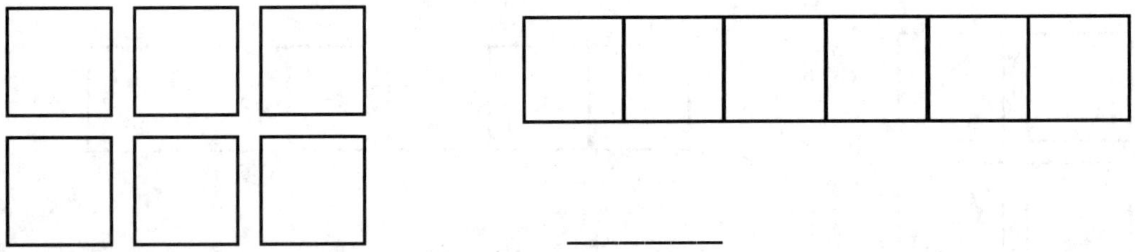

_____

_____

_____

_____

SYSTEMATIC REVIEW

Count, match, and color. Write the number and say it.

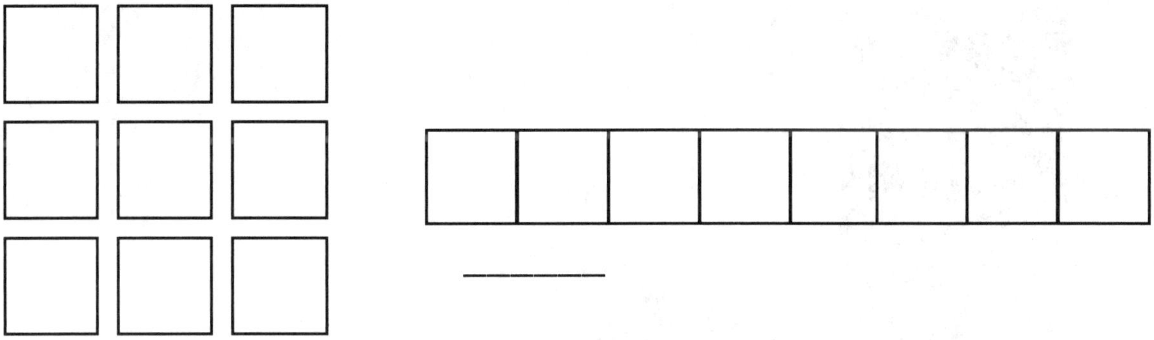

_____

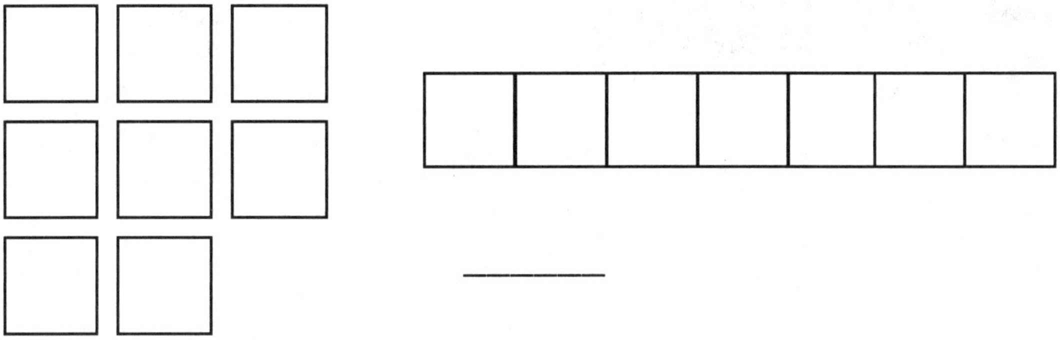

_____

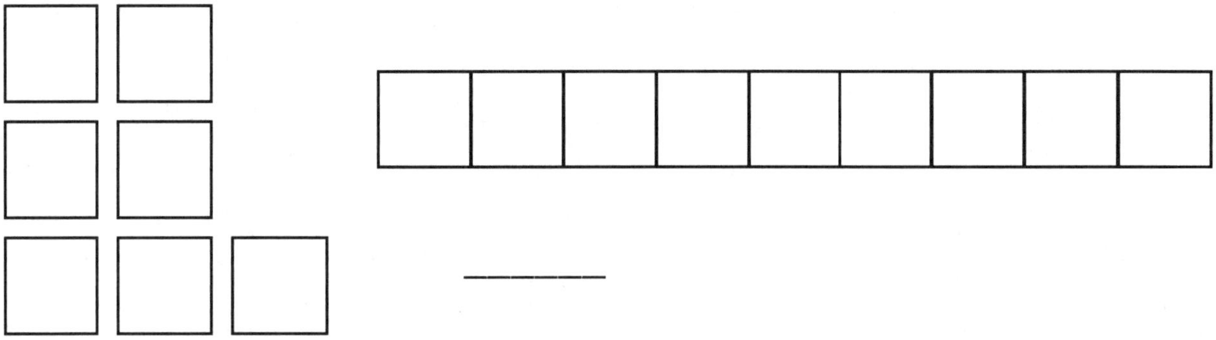

_____

Count and write. Say the number.

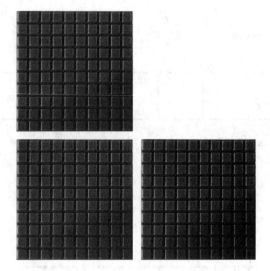

_____   _____   _____

SYSTEMATIC REVIEW

Count, match, and color. Write the number and say it.

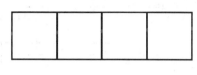

_____

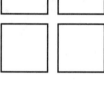

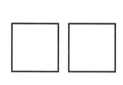

_____

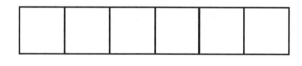

_____

Color the correct number of blocks. Say the number.

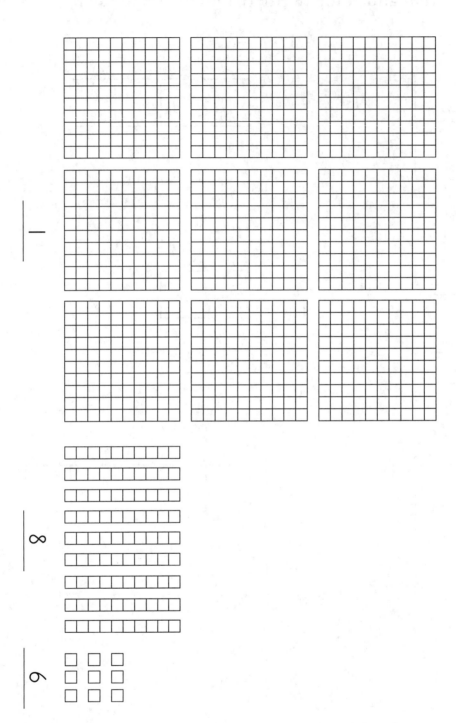

11F

Count, match, and color. Write the number and say it.

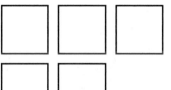

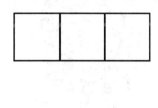

Count and write. Say the number.

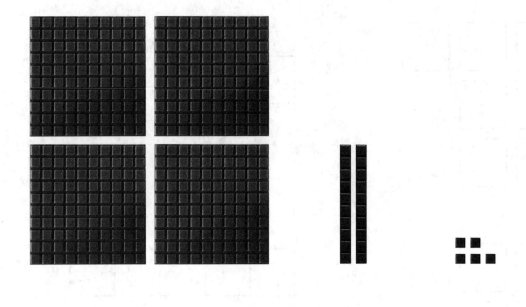

_____   _____   _____

APPLICATION AND ENRICHMENT

Draw lines to match the blocks with the pictures. The first one has been done for you. Color the pictures to match the Math-U-See blocks.

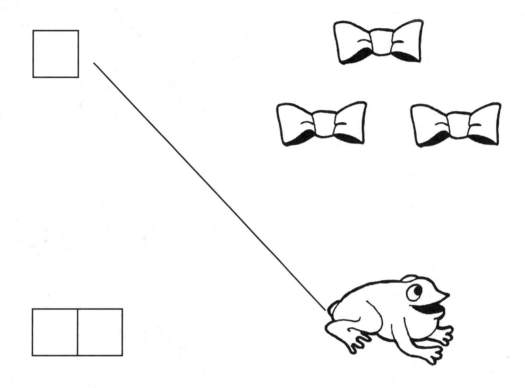

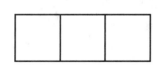

Draw lines to match the blocks with the pictures. Color the pictures to match the Math-U-See blocks.

LESSON PRACTICE

12A

Build each addition problem. Say it and write it. Color the picture to match the unit bars. Turn the book sideways first.

	+	+

| | = | = |

12B

Build each addition problem. Say it and write it. Color the
picture to match the unit bars.

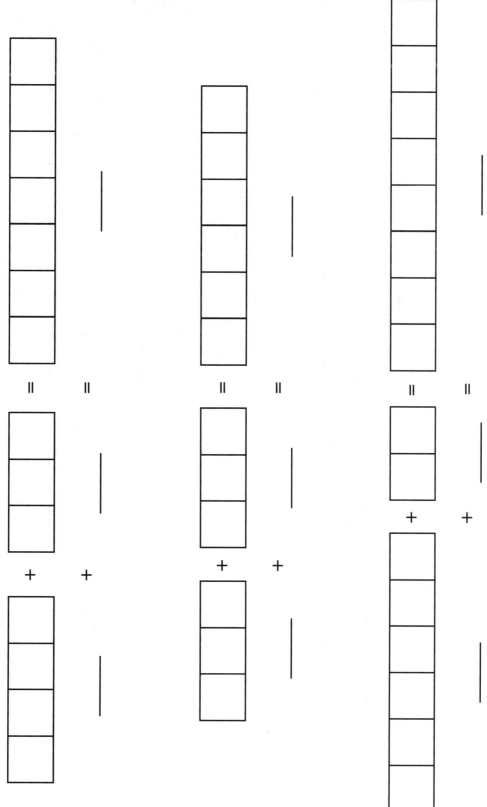

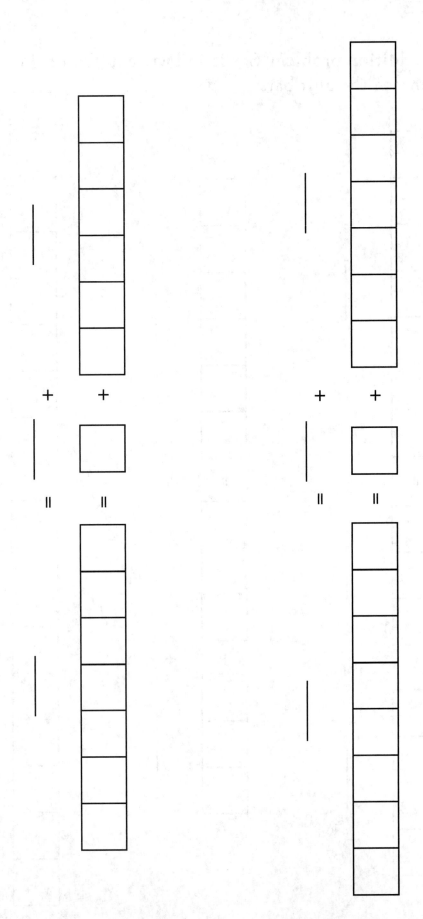

LESSON PRACTICE

Build each addition problem. Say it and write it. Color the picture to match the unit bars.

Build each addition problem. Say it and write it.
Color the picture to match the unit bars.

Count and write. Say the number.

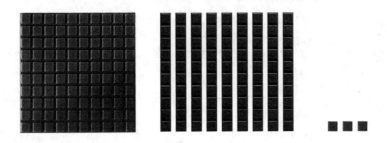

_____   _____   _____

Count the rectangles. Circle and say the correct number. ⬚

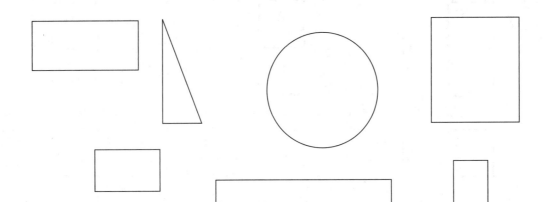

0 1 2 3 4 5 6 7 8 9

12E

Build each addition problem. Say it and write it. Color the picture to match the unit bars.

Color the correct number of blocks. Say the number.

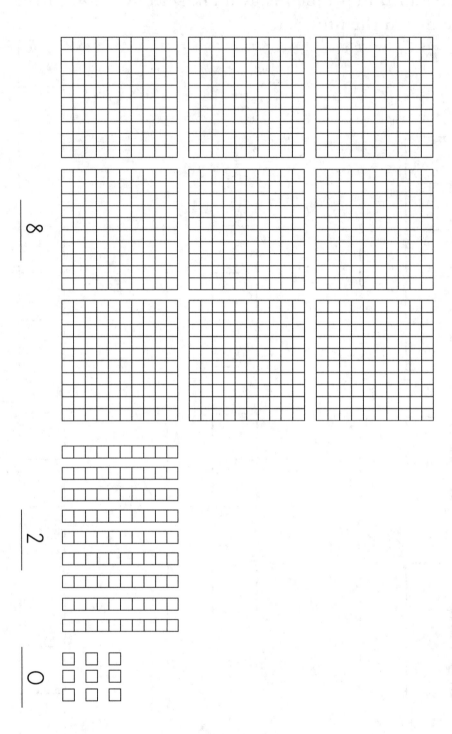

Build each addition problem. Say it and write it. Color the picture to match the unit bars.

Count and write. Say the number.

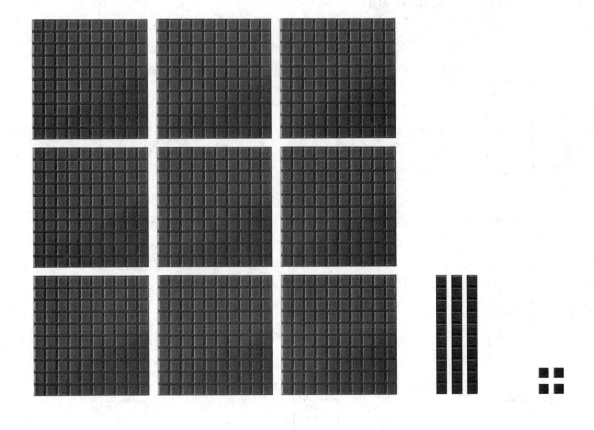

——————   ——————   ——————

Match the blocks with the pictures. Color the pictures to match the Math-U-See blocks.

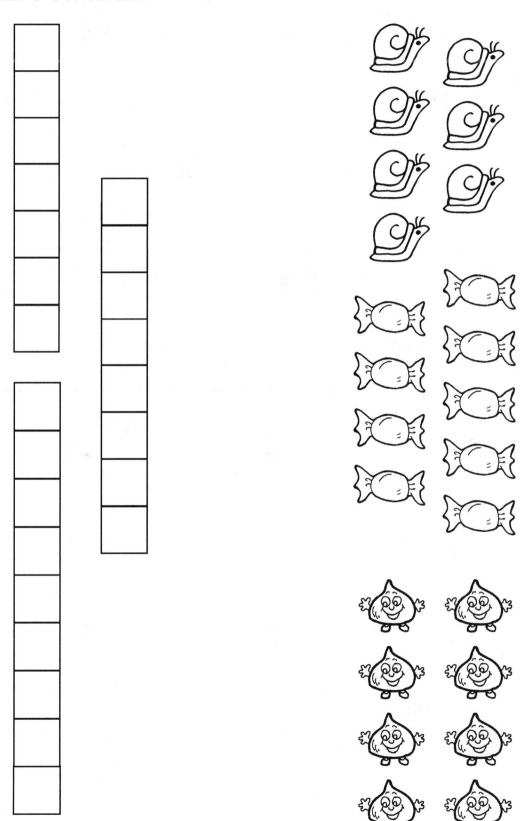

Dot-to-dot activities are a great way to practice counting. Start at 1 and connect the dots to draw the shapes.

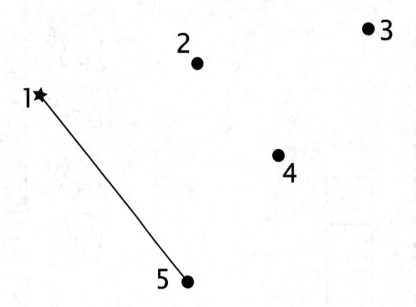

What shape is this?

_____

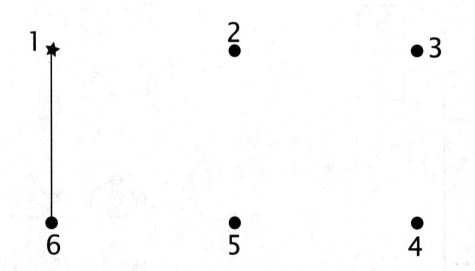

What shape is this?

LESSON PRACTICE

Build, say, write, and color.

$\square$ + $\square$ = $\square\square$

___ + ___ = ___

$\square\square$ + $\square$ = $\square\square\square$

___ + ___ = ___

$\square\square\square\square$ + $\square$ = $\square\square\square\square\square$

___ + ___ = ___

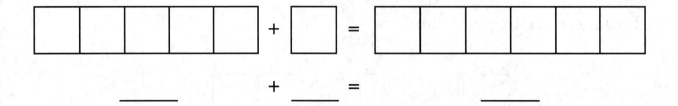

$$\square\square\square\square\square + \square = \square\square\square\square\square\square$$

$$\underline{\quad} + \underline{\quad} = \qquad \underline{\quad}$$

$$\square\square\square + \square = \square\square\square\square$$

$$\underline{\quad} + \underline{\quad} = \qquad \underline{\quad}$$

LESSON PRACTICE

Build, say, write, and color.

‖   ‖   ‖   ‖   ‖   ‖

+   +   +   +   +   +

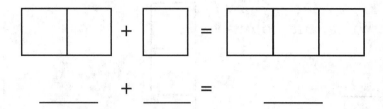

$$ \rule{3em}{0.4pt} + \rule{3em}{0.4pt} = \rule{3em}{0.4pt} $$

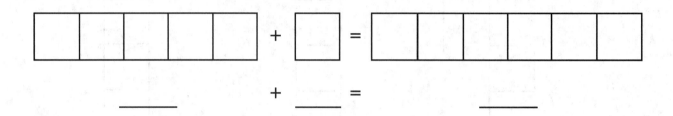

$$ \rule{3em}{0.4pt} + \rule{3em}{0.4pt} = \rule{3em}{0.4pt} $$

LESSON PRACTICE

Build, say, write, and color.

☐☐☐☐ + ☐ = ☐☐☐☐☐

___ + ___ = ___

☐ + ☐ = ☐☐

___ + ___ = ___

☐☐☐ + ☐ = ☐☐☐☐

___ + ___ = ___

Build, say, write, and color.

Count and write. Say the number.

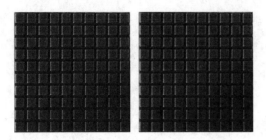

_____   _____   _____

Count the circles. Circle and say the correct number. ◯

0 1 2 3 4 5 6 7 8 9

SYSTEMATIC REVIEW

Build, say, write, and color.

$$\boxed{||} + \boxed{} = \boxed{|||}$$

___ + ___ = ___

$$\boxed{} + \boxed{} = \boxed{|}$$

___ + ___ = ___

$$\boxed{|||} + \boxed{} = \boxed{||||}$$

___ + ___ = ___

Color the correct number of blocks. Say the number.

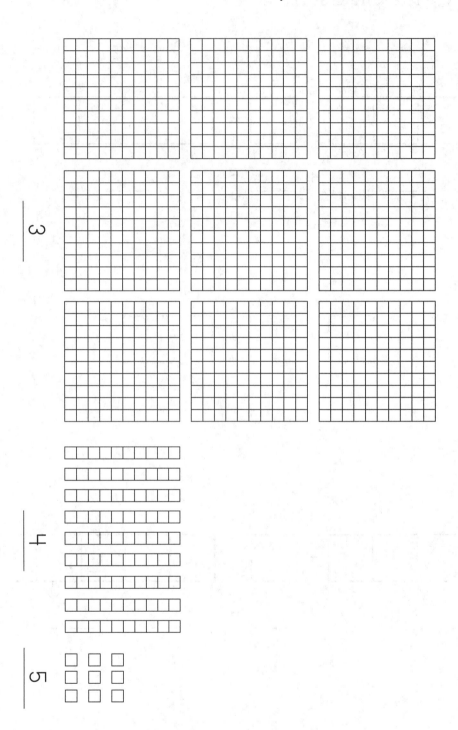

Build, say, write, and color.

Count and write. Say the number.

———   ———   ———

Count the triangles. Circle and say the correct number. △

0 1 2 3 4 5 6 7 8 9

Add. Write the answer on the line.

1 squirrel   +   1 squirrel   = _____ squirrels

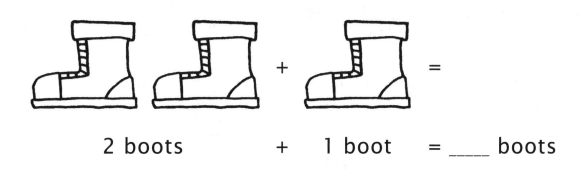

2 boots   +   1 boot   = _____ boots

3 flowers   +1 flower   = _____ flowers

Start at 1 and connect the dots to finish the picture.

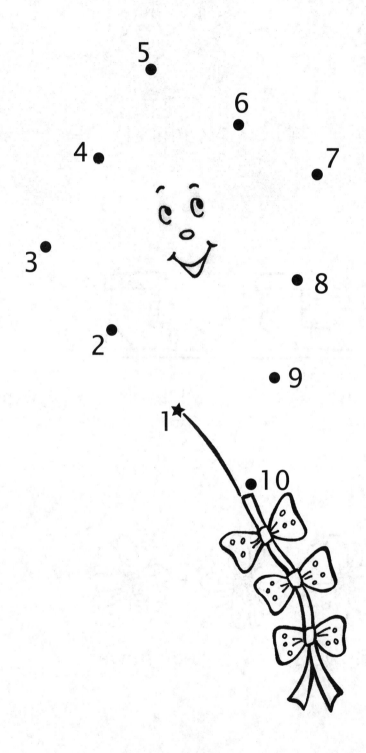

14A

Write the numbers that match the blocks. Some numbers have already been written for you.

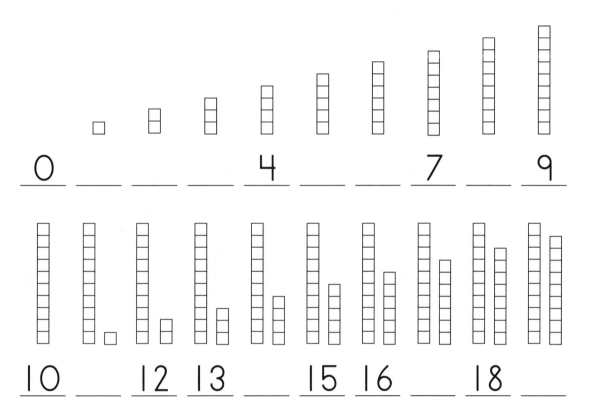

0 ___ ___ ___ 4 ___ ___ 7 ___ 9

10 ___ 12 13 ___ 15 16 ___ 18 ___

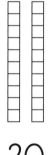

20

Count and write from zero to twenty.

0 ___ 2 ___ ___ 5 ___ ___ 8 ___

___ 11 ___ 13 ___ ___ ___ 17 ___ 19

___

Write the numbers that match the blocks. Some numbers have already been written for you.

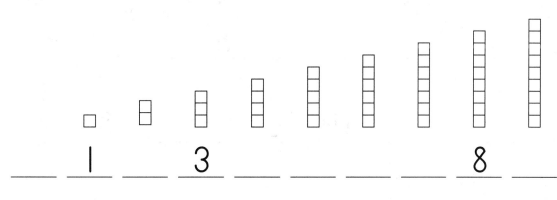

___ 1 ___ 3 ___ ___ ___ ___ 8 ___

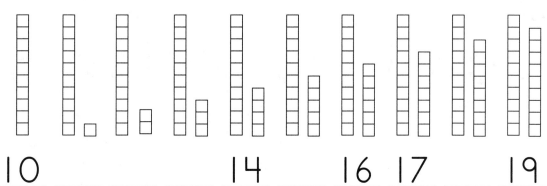

10 ___ ___ ___ 14 ___ 16 17 ___ 19

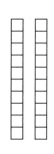

___

Count and write from zero to twenty.

$\underline{0}$ $\underline{}$ $\underline{2}$ $\underline{}$ $\underline{}$ $\underline{}$ $\underline{6}$ $\underline{}$ $\underline{}$ $\underline{9}$

$\underline{}$ $\underline{11}$ $\underline{}$ $\underline{}$ $\underline{14}$ $\underline{15}$ $\underline{}$ $\underline{}$ $\underline{18}$ $\underline{}$

$\underline{20}$

Write the numbers that match the blocks. Some numbers have already been written for you.

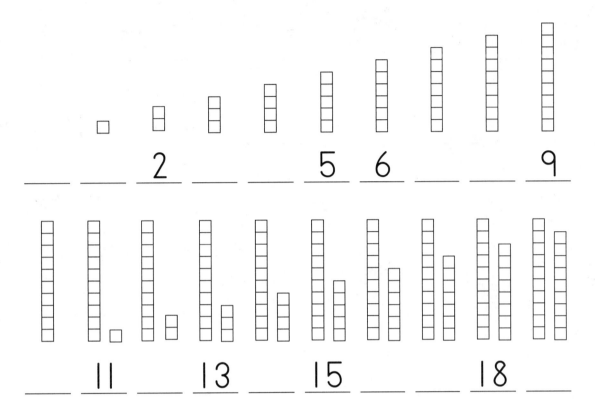

___ ___ 2 ___ ___ 5 6 ___ ___ 9

___ 11 ___ 13 ___ 15 ___ ___ 18 ___

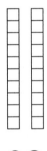

20

Count and write from zero to twenty.

0 __ 2 __ __ 5 __ __ 8 __

__ 11 __ 13 __ __ __ 17 __ 19

__

Count and write from zero to twenty.

$\underline{0}$ __ __ __ $\underline{4}$ __ __ __ $\underline{8}$ __

__ $\underline{11}$ $\underline{12}$ __ __ __ $\underline{16}$ __ __ __

$\underline{20}$

....

Build, color, match, and write. The first one has been done for you.

$\underline{8} + \underline{1} =$

_____

$\underline{3} + \underline{1} =$

$\underline{9}$

$\underline{6} + \underline{1} =$

_____

Count and write. Say the number.

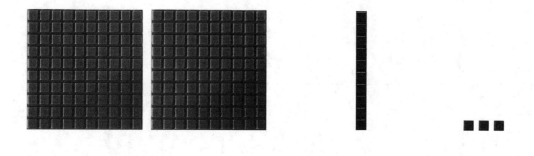

_____   _____   _____

Count and write from zero to twenty.

__ 1 __ __ __ 5 6 __ __ __

10 __ __ 13 14 __ __ 17 __ 19

__

Build, color, match, and write.

1 + 1 =

___

7 + 1 =

___

5 + 1 =

___

Count and write. Say the number.

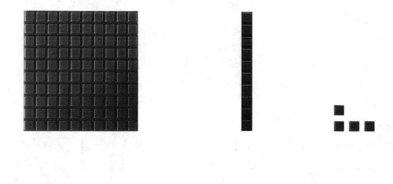

_____   _____   _____

14F

Count and write from zero to twenty.

___ 2 ___ 4 ___ ___ ___ 9 ___

___ ___ ___ ___ 15 ___ ___ 18 ___

20

Build, color, match, and write.

4 + 1 =

_____

2 + 1 =

_____

8 + 1 =

_____

Count and write. Say the number.

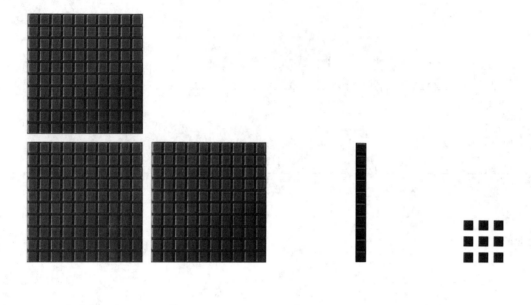

_____   _____   _____

14G

The student has been using unit, ten, and hundred blocks to build numbers. This activity asks them to start with unit blocks only. Have the student count out ten of the units and exchange them for a ten bar, and then discover how many unit blocks are left over. Try the example given here.

Help the student count out thirteen unit blocks. How many tens and units are the same as thirteen units?

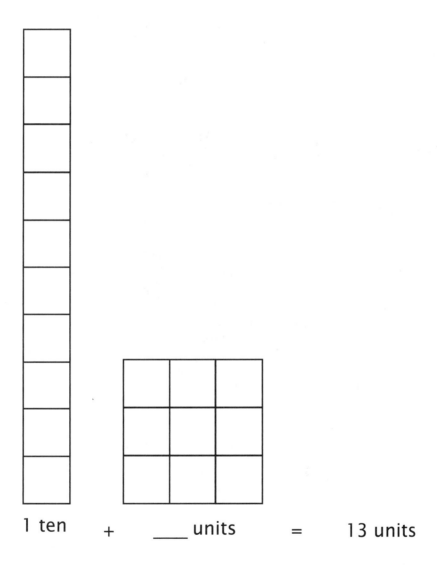

1 ten   +   ____ units   =   13 units

Help your student build other numbers greater than ten by starting with the unit blocks and talking through the process. Write and say the results.

Start at 1 and connect the dots to finish the picture.

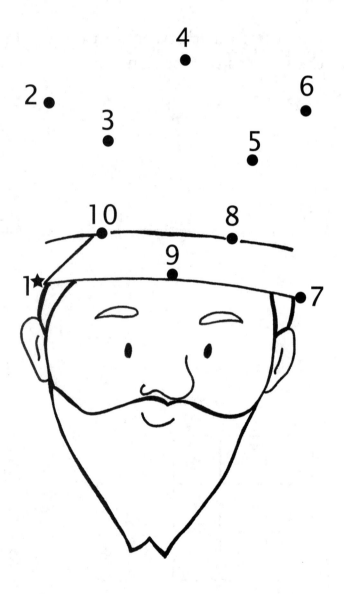

LESSON PRACTICE

Build, say, write, and color.

$$\boxed{|} \; + \; \boxed{|} \; = \; \boxed{|||}$$

____ + ____ = ____

$$\boxed{||} \; + \; \boxed{||} \; = \; \boxed{|||||}$$

____ + ____ = ____

Build, color, match, and write.

$\underline{2} + \underline{2} =$ $\boxed{|||||}$

____

$\underline{6} + \underline{1} =$ $\boxed{||||||}$

____

$\underline{3} + \underline{3} =$ $\boxed{|||}$

____

Build and write. The first one has been done for you.

$$\begin{array}{r} 3 \\ +\ \ 3 \\ \hline 6 \end{array}$$

$$\begin{array}{r} 2 \\ +\ \ 1 \\ \hline \end{array}$$

$$\begin{array}{r} 2 \\ +\ \ 2 \\ \hline \end{array}$$

$$\begin{array}{r} 1 \\ +\ \ 5 \\ \hline \end{array}$$

15B

Build, say, write, and color.

☐☐☐ + ☐☐☐ = ☐☐☐☐☐☐

_____ + _____ = _____

☐☐ + ☐☐ = ☐☐☐☐

_____ + _____ = _____

....

Build, color, match, and write.

_7_ + _1_ =   ☐☐☐☐

_____

_3_ + _3_ =   ☐☐☐☐☐☐☐☐

_____

_2_ + _2_ =   ☐☐☐☐☐☐

_____

Build and write.

```
      3              2
  +   1          +   2
  _____        _____


      3              1
  +   3          +   1
  _____        _____
```

Build, say, write, and color.

☐☐ + ☐☐ = ☐☐☐☐

_____ + _____ = _____

☐☐☐ + ☐☐☐ = ☐☐☐☐☐☐

_____ + _____ = _____

....

Build, color, match, and write.

$\underline{3} + \underline{3} =$ ☐☐☐☐☐☐

_____

$\underline{2} + \underline{2} =$ ☐☐☐☐☐☐☐☐☐

_____

$\underline{8} + \underline{1} =$ ☐☐☐☐

_____

Build and write.

$$\begin{array}{r} 2 \\ +\ 2 \\ \hline \end{array} \qquad\qquad \begin{array}{r} 4 \\ +\ 1 \\ \hline \end{array}$$

$$\begin{array}{r} 1 \\ +\ 2 \\ \hline \end{array} \qquad\qquad \begin{array}{r} 3 \\ +\ 3 \\ \hline \end{array}$$

Build, color, and write.

$$\boxed{}\boxed{}\boxed{} \; + \; \boxed{}\boxed{}\boxed{}$$

$$\underline{} \; + \; \underline{} \; = \; \underline{}$$

Build and write.

$$\begin{array}{r} 5 \\ + \; 1 \\ \hline \end{array} \qquad\qquad \begin{array}{r} 2 \\ + \; 2 \\ \hline \end{array}$$

$$\begin{array}{r} 1 \\ + \; 6 \\ \hline \end{array} \qquad\qquad \begin{array}{r} 1 \\ + \; 1 \\ \hline \end{array}$$

Count and write from zero to twenty.

0 __ __ __ __ 5 __ __ __ 9

__ 11 __ 13 __ __ __ __ 18 __

20

....

Count and write the number. Then say it.

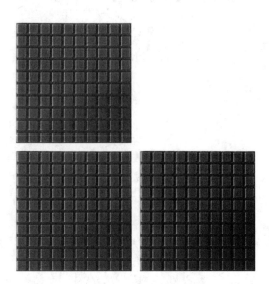

____   ____   ____

Build, color, and write.

☐☐ + ☐☐

____ + ____ = ____

Build and write.

```
    3              7
+   3          +   1
_____         _____
```

```
    1              4
+   3          +   1
_____         _____
```

Count and write from zero to twenty.

___ 1 ___ ___ 4 ___ ___ ___ 8 ___

10 ___ 12 ___ ___ ___ ___ ___ 19

___

Count and write. Say the number.

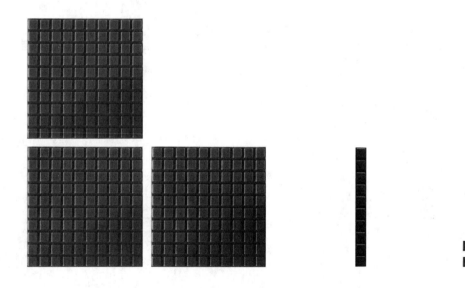

___ ___ ___

15F

SYSTEMATIC REVIEW

Build and write.

$$\begin{array}{r} 8 \\ +\ 1 \\ \hline \end{array} \qquad \begin{array}{r} 2 \\ +\ 2 \\ \hline \end{array}$$

$$\begin{array}{r} 3 \\ +\ 3 \\ \hline \end{array} \qquad \begin{array}{r} 2 \\ +\ 1 \\ \hline \end{array}$$

Count and write from zero to twenty.

___ ___ 2 3 ___ 5 ___ 7 ___ ___

___ ___ ___ ___ 14 ___ ___ 17 ___ ___

20

Count and write. Say the number.

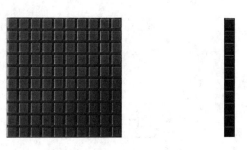

_____   _____   _____

Build and say the numbers.

412

321

Add. Write the answer on the line.

2 trees   +   2 trees   = _____ trees

3 ducks   +   3 ducks   = _____ ducks

2 apples   +   2 apples   = _____ apples

Start at 1 and connect the dots to finish the picture.

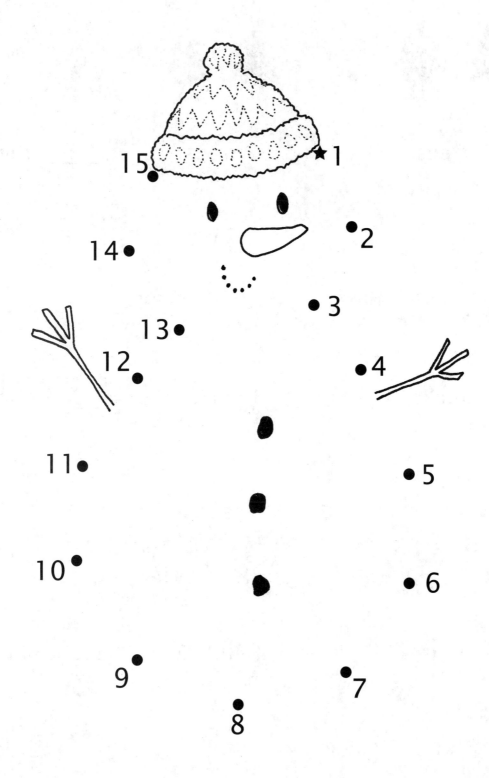

The diagrams are smaller than actual size. Have the student count to select the correct unit bars to complete the problems. Build, say, write, and color.

| | | | | | + | | | | | = | | | | | | | | | |

_____ + _____ = _____

| | | | | | + | | | | | = | | | | | | | | | | |

_____ + _____ = _____

Build and write.

```
      5              4
  +   5          +   4
  _____        _____


      1              2
  +   6          +   2
  _____        _____
```

Count the squares that match the pictures. Circle and say the correct number.

....

How many squares are white? ☐

....

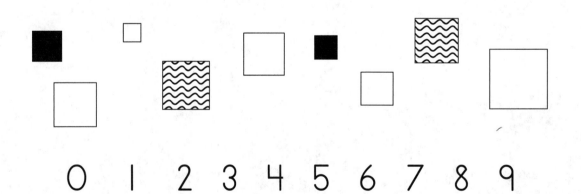

0   1   2   3   4   5   6   7   8   9

The diagrams are smaller than actual size. Have the student count to select the correct unit bars to complete the problems. Build, say, write, and color.

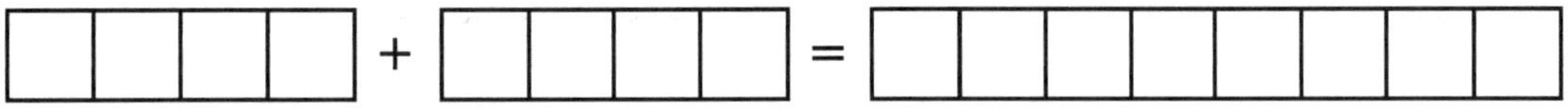

_____ + _____ = _____

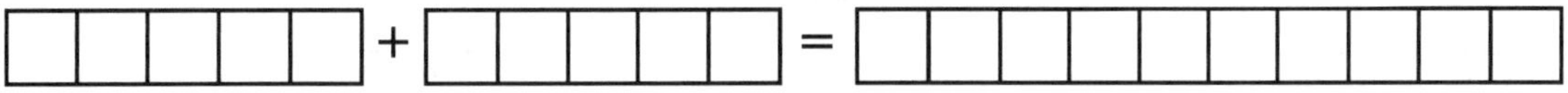

_____ + _____ = _____

Build and write.

$$\begin{array}{r} 4 \\ +\ 4 \\ \hline \end{array} \qquad \begin{array}{r} 3 \\ +\ 3 \\ \hline \end{array}$$

$$\begin{array}{r} 5 \\ +\ 5 \\ \hline \end{array} \qquad \begin{array}{r} 5 \\ +\ 1 \\ \hline \end{array}$$

Count the squares that match the pictures. Circle and say the correct number.

How many squares are black?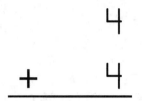

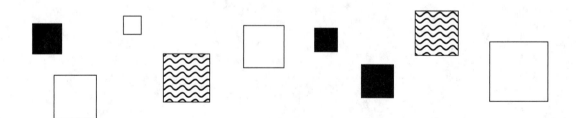

0 1 2 3 4 5 6 7 8 9

Build, color, match, and write.

$\underline{4} + \underline{4} =$

———

$\underline{1} + \underline{1} =$

———

$\underline{5} + \underline{5} =$

———

Build and write.

$$\begin{array}{r} 1 \\ +\ 3 \\ \hline \end{array}$$

$$\begin{array}{r} 7 \\ +\ 1 \\ \hline \end{array}$$

$$\begin{array}{r} 2 \\ +\ 2 \\ \hline \end{array}$$

$$\begin{array}{r} 1 \\ +\ 4 \\ \hline \end{array}$$

Count the squares that match the pictures. Circle and say the correct number.

How many squares have waves?

0 1 2 3 4 5 6 7 8 9

Build and write.

$$
\begin{array}{r}
1 \\
+\ 8 \\
\hline
\end{array}
\qquad
\begin{array}{r}
4 \\
+\ 4 \\
\hline
\end{array}
$$

$$
\begin{array}{r}
2 \\
+\ 1 \\
\hline
\end{array}
\qquad
\begin{array}{r}
5 \\
+\ 5 \\
\hline
\end{array}
$$

Count the squares that match the pictures. Circle and say the correct number.

How many squares are gray?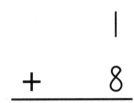

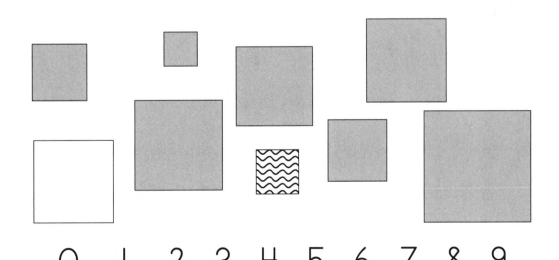

0 1 2 3 4 5 6 7 8 9

Count and write from zero to twenty.

0 __ __ __ __ __ 6 __ __ __

__ 11 __ __ __ __ __ __ 19

20

....

Build and say the numbers.

216

161

Build and write.

```
    3              6
+   3          +   1
_____        _____
```

```
    5              4
+   5          +   4
_____        _____
```

Count the squares that match the pictures. Circle and say the correct number.

How many squares have rectangles in them? ▦

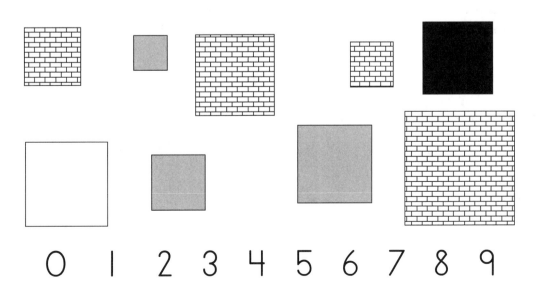

0 1 2 3 4 5 6 7 8 9

Count and write from zero to twenty.

_____ 2 _____ _____ _____ 7 _____ _____

_____ _____ _____ 15 _____ 18 _____

_____

Build and say the numbers.

318

410

Build and write.

$$\begin{array}{r} 5 \\ + 5 \\ \hline \end{array}$$

$$\begin{array}{r} 5 \\ + 1 \\ \hline \end{array}$$

$$\begin{array}{r} 4 \\ + 4 \\ \hline \end{array}$$

$$\begin{array}{r} 1 \\ + 3 \\ \hline \end{array}$$

Count the squares that match the pictures. Circle and say the correct number.

How many squares have waves?

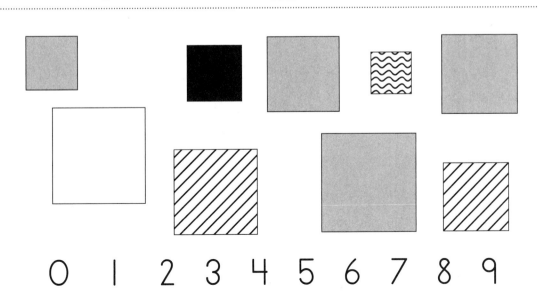

0 1 2 3 4 5 6 7 8 9

Count and write from zero to twenty.

__0__ ___ ___ ___ ___ __5__ ___ ___ ___ ___

___ __11__ ___ ___ ___ ___ ___ __17__ ___ ___

__20__

Build and say the numbers.

III

231

16G

Follow the directions.

If the square has a 1, color it red.
If the square has a 2, color it green.
Color the last square to match the pattern.

1	2	1	2	1	

If the square has a 1, color it blue.
If the square has a 2, color it yellow.
Color the last square to match the pattern.

1	2	2	1	2	

If the square has a 1, color it orange.
If the square has a 2, color it brown.
Color the last two squares to match the pattern.

2	2	1	2	2		

Start at 1 and connect the dots to finish the picture.

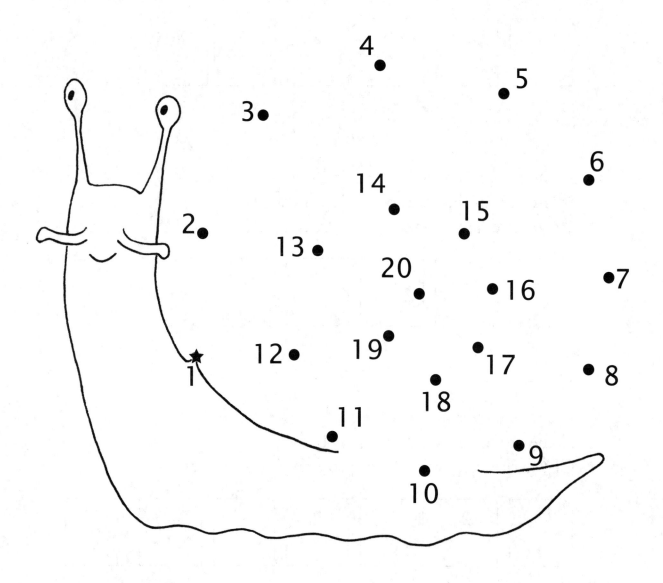

How many squares are there? Skip count by two and write the
numbers on the lines. Write the number of squares on the line
outside the boxes. The first one has been done for you.

	2
	4
	6
	8
	10

____10____

	2
	___
	6
	___
	___
	12

_____

	____
	_4_
	____
	____
	_10_
	____
	_14_

	_2_
	____
	____
	_8_
	____
	_12_
	____

_____   _____

....

Skip count by two to find how many circles.

○   ○   ○   ○   ○
○   ○   ○   ○   ○

_____ circles

LESSON PRACTICE

How many squares are there? Skip count by two. Write the numbers on the lines.

	2
	___
	___
	___
	10
	___
	___
	16

	___
	4
	___
	8
	___
	12
	___
	18

_____   _____

Skip count by two to find how many rectangles.

_____ rectangles

Skip count by two to find how many candies.

_____ candies

17C

How many squares are there? Skip count by two. Write the numbers on the lines.

	____
	____
	_6_
	____
	____
	____
	_14_
	____
	_18_

	_2_
	____
	____
	____
	_10_
	____
	____
	_16_
	____
	_20_

_____

_____

Skip count by two to find how many triangles.

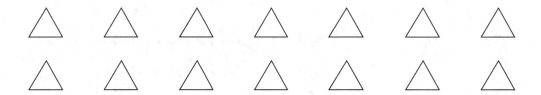

_____ triangles

Skip count by two to find how many houses.

_____ houses

SYSTEMATIC REVIEW

How many squares are there? Skip count by two. Write the numbers on the lines.

	____
	4
	____
	____
	____
	12
	____
	16
	____

	____
	____
	6
	____
	____
	____
	14
	____
	18
	____

_____

_____

Skip count by two to find how many chairs.

_____ chairs

Build and write.

$$\begin{array}{r} 7 \\ +\ 1 \\ \hline \end{array}$$

$$\begin{array}{r} 1 \\ +\ 4 \\ \hline \end{array}$$

$$\begin{array}{r} 2 \\ +\ 2 \\ \hline \end{array}$$

$$\begin{array}{r} 1 \\ +\ 1 \\ \hline \end{array}$$

SYSTEMATIC REVIEW

How many squares are there? Skip count by two. Write the numbers on the lines.

	2
	___
	___
	___
	10
	___

	___
	___
	___
	8
	___
	12
	___

_____   _____

Skip count by two to find how many butterflies.

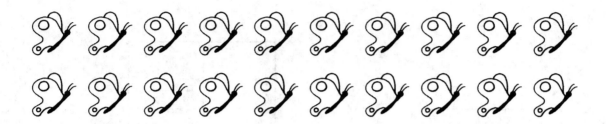

_____ butterflies

Build and write.

2 + 1 = _____   1 + 8 = _____

```
    3              6
+   3          +   1
-----          -----
```

```
    5              1
+   5          +   3
-----          -----
```

SYSTEMATIC REVIEW

How many squares are there? Skip count by two. Write the
numbers on the lines.

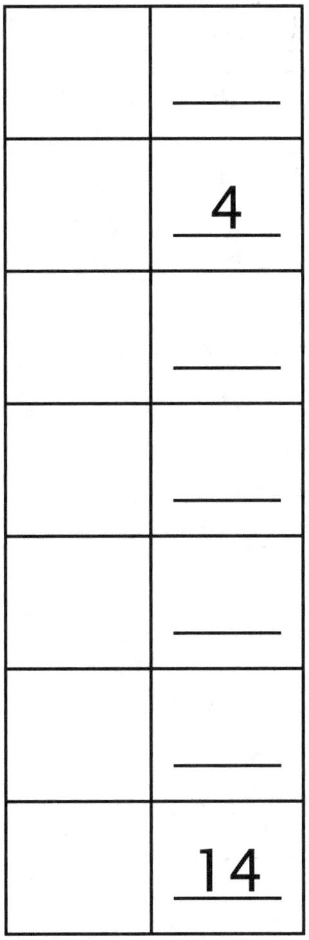

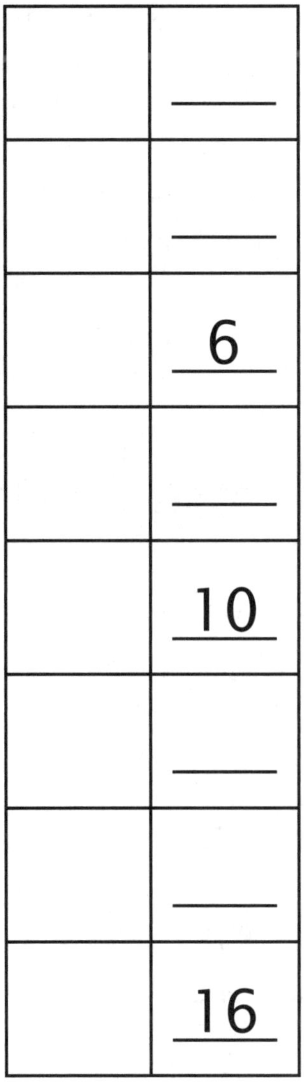

Left grid (top to bottom): ___, 4, ___, ___, ___, ___, 14

Right grid (top to bottom): ___, ___, 6, ___, 10, ___, ___, 16

___   ___

Skip count by two to find how many fish.

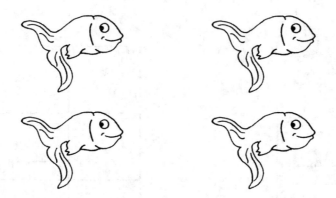

_____ fish

Build and write.

$4 + 4 =$ ___ $4 + 1 =$ ___

$$\begin{array}{r} 5 \\ + \ 5 \\ \hline \end{array}$$ $$\begin{array}{r} 1 \\ + \ 7 \\ \hline \end{array}$$

$$\begin{array}{r} 2 \\ + \ 2 \\ \hline \end{array}$$ $$\begin{array}{r} 1 \\ + \ 1 \\ \hline \end{array}$$

Begin at the star by the number 2. Skip count by two and connect the dots to finish the picture.

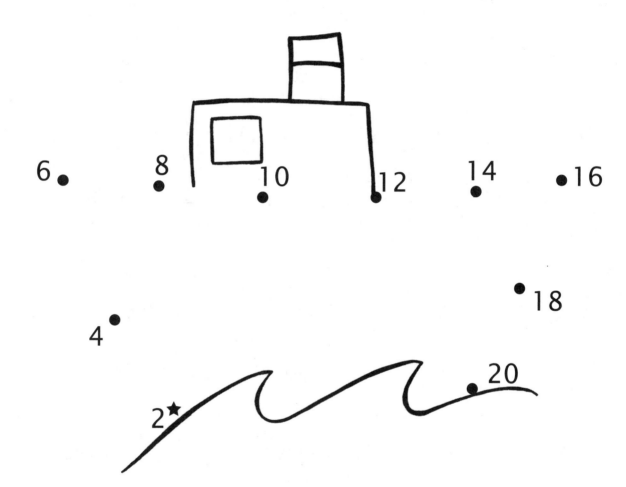

Add. Write the answer on the line.

4 chairs   +   4 chairs   = _____ chairs

3 fish   +   3 fish   = _____ fish

5 leaves   +   5 leaves   = _____ leaves

Build and add the tens. Say the problem as you write it.

$$\underline{10} + \underline{10} = \underline{}$$

$$\underline{} + \underline{} = \underline{}$$

$$\underline{} + \underline{} = \underline{}$$

Build and add the tens. Write the answer.

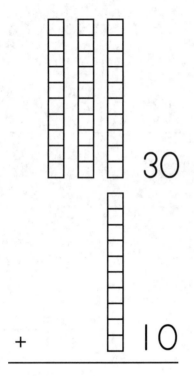

30

+ 10

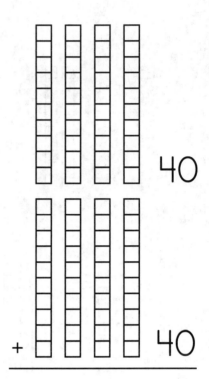

40

+ 40

Build and add the tens. Say the problem as you write it.

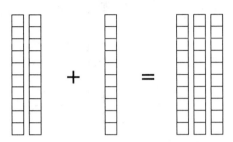

_____ + _____ = _____

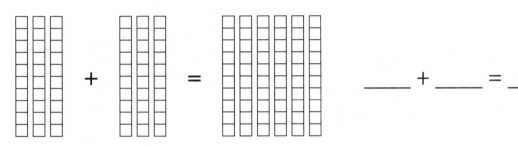

_____ + _____ = _____

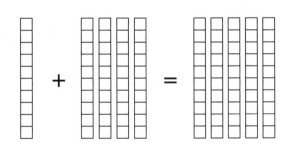

_____ + _____ = _____

Build and add the tens. Write the answer.

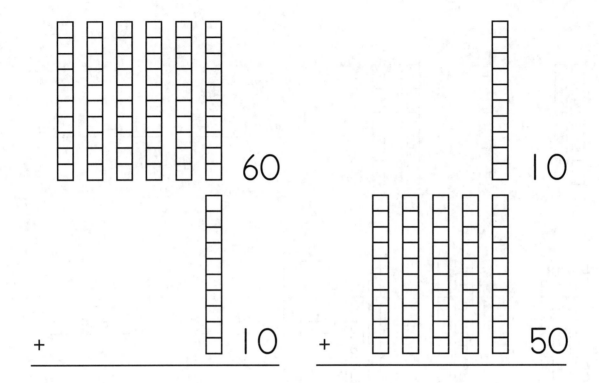

LESSON PRACTICE

Build and add the tens. Say the problem as you write it.

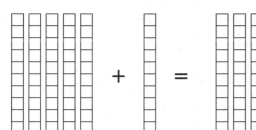

___ + ___ = ___

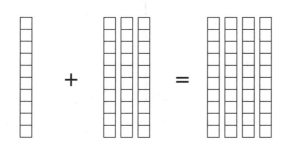

___ + ___ = ___

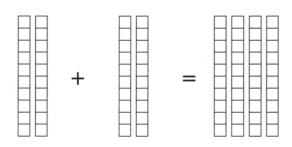

___ + ___ = ___

Build and add the tens. Write the answer.

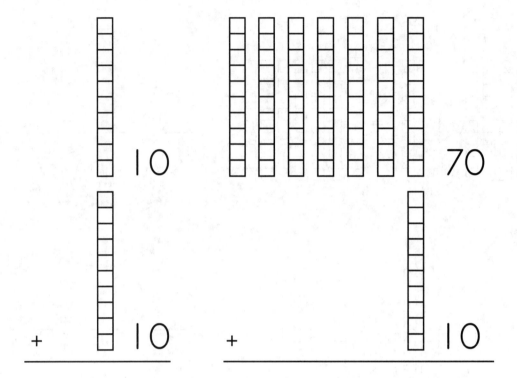

SYSTEMATIC REVIEW

Build and add the tens. Write and say the answer.

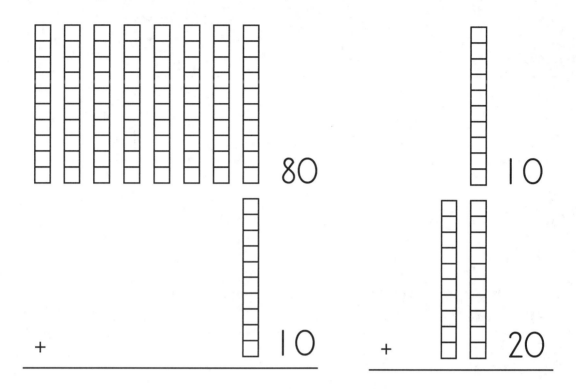

80
+ 10
_____

10
+ 20
_____

Skip count by two to find how many balloons.

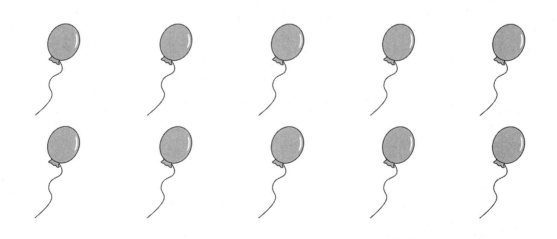

_____ balloons

Count and write from zero to twenty.

___ ___ 2 ___ 4 ___ ___ ___ ___ 9 ___

___ ___ ___ ___ ___ ___ ___ ___ 18 ___

20 ___

Build and add the tens. Write and say the answer.

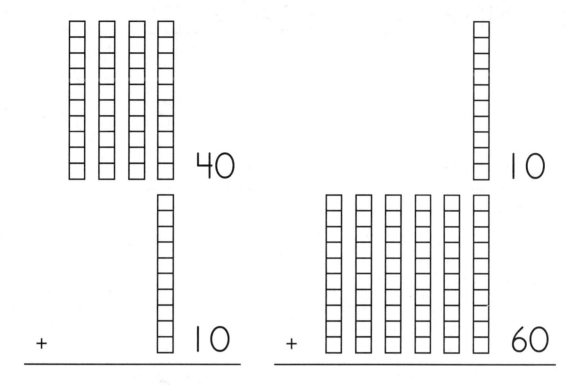

40

+ 10

10

+ 60

Skip count by two to find how many circles.

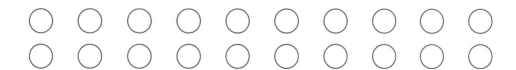

_____ circles

Build and write.

$$40 + 40 = \underline{\quad} \qquad 2 + 1 = \underline{\quad}$$

$$\begin{array}{r} 1 \\ + 5 \\ \hline \end{array} \qquad \begin{array}{r} 3 \\ + 3 \\ \hline \end{array}$$

$$\begin{array}{r} 6 \\ + 1 \\ \hline \end{array} \qquad \begin{array}{r} 70 \\ + 10 \\ \hline \end{array}$$

Build and add the tens. Write and say the answer.

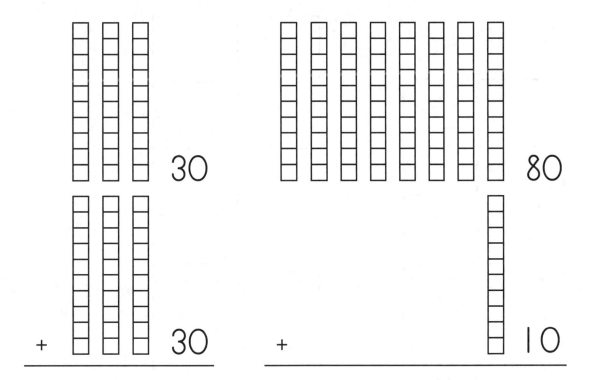

Build and write.

$10 + 30 =$ _____

$20 + 20 =$ _____

How many squares are there? Skip count by two. Write the numbers on the lines.

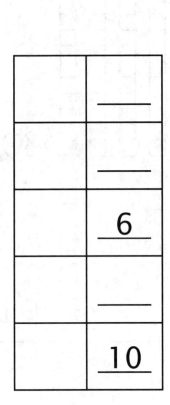

Build and say the numbers.

416

324

Begin at the star by the number 2. Skip count by two and connect the dots to finish the picture.

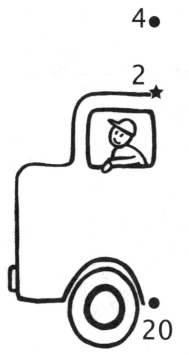

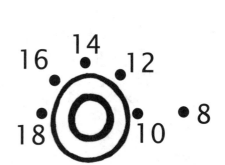

Add.

If the answer is 40, color the space blue.
If the answer is 60, color the space green.

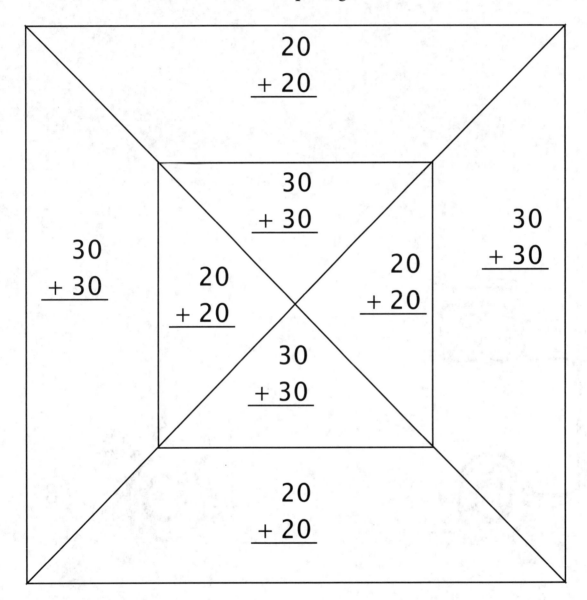

What shapes do you see?

To the parent: Your child should easily see the two common shapes that have been
introduced so far. The shapes in the outside border are called trapezoids.

Skip count by 10 and write the numbers on the lines. Then write the numbers in the spaces under the squares.

								10
								___
								30
								___
								___

___ , 20, ___ , ___ , 50

Skip count by 10 and write the numbers on the lines. Then write the numbers in the spaces under the squares.

									___
									_20_
									___
									___
									___
									_60_
									___
									_80_
									___
									_100_

10, __ ,30, __ , __ , __ ,70, __ ,90, 100

LESSON PRACTICE

Skip count by 10 and write the numbers on the lines. Then write the numbers in the spaces under the squares.

								___
								<u>20</u>
								___
								___

10, ___ , ___ , ___

Skip count by 10 and write the numbers on the lines. Then write the numbers in the spaces under the squares.

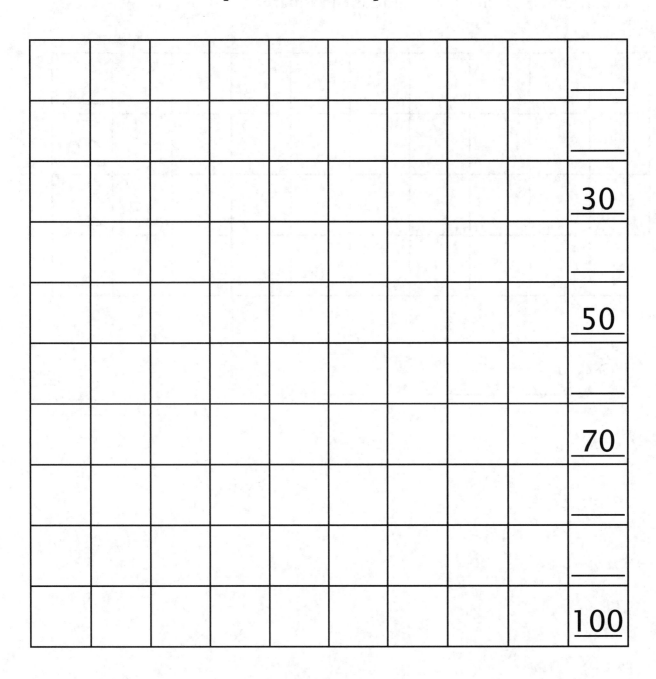

									____
									____
									30
									____
									50
									____
									70
									____
									____
									100

__ , 20, __ , __ , __ , 60, __ , 80, __ , __

Skip count by 10 and write the numbers on the lines. Then write the numbers in the spaces under the squares.

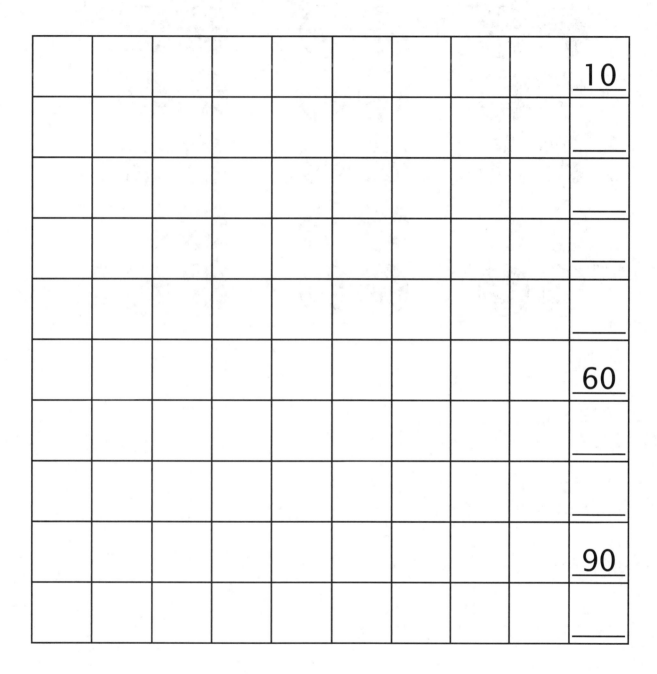

__, __ ,30, __ , __ , __ ,70, __ , __ ,100

Skip count by 10 to find how many marbles.

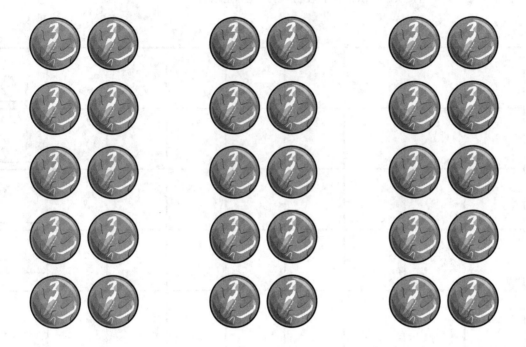

_____ marbles

Skip count by 10. Write the missing numbers on the lines.

10, __ , __ , __ ,50, __ , __ , __ ,90, ___

Skip count by 10 to find how many squares.

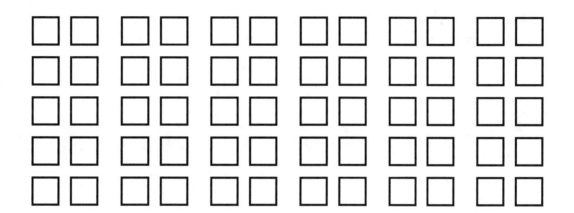

_____ squares

Build and write.

$10 + 10 = \underline{}$ $4 + 1 = \underline{}$

$$\begin{array}{r} 1 \\ + 7 \\ \hline \end{array} \qquad\qquad \begin{array}{r} 2 \\ + 2 \\ \hline \end{array}$$

$$\begin{array}{r} 2\,0 \\ + 1\,0 \\ \hline \end{array} \qquad\qquad \begin{array}{r} 4\,0 \\ + 4\,0 \\ \hline \end{array}$$

Read the word problem to the student if necessary. Use the pictures to help solve it.

Pam has three pieces of cake. Pat has one piece of cake. How many pieces of cake do Pam and Pat have in all?

+ = \underline{} pieces

Skip count by 10. Write the missing numbers on the lines.

__ ,20, __ , __ , __ ,60, __ ,80, __ , ____

Skip count by 2. Write the missing numbers on the lines.

2, __ ,6, __ , __ ,12, __ , __ , __ ,20

Skip count by 10 to find how many leaves.

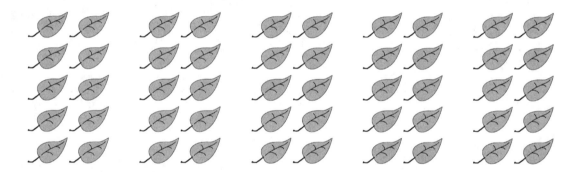

_____ leaves

Build and write.

```
    4            8
 +  4         +  1
 _____       _____


   1 0          7 0
 + 5 0        + 1 0
 _____       _____
```

Read the word problem to the student if necessary. Use the pictures to help solve it.

Tom had one pencil. Papa gave Tom one more pencil. How many pencils does Tom have now?

 + = _____ pencils

Skip count by 10. Write the missing numbers on the lines.

10, __ , __ , __ ,50, __ , __ , __ , __ ,100

Skip count by 2. Write the missing numbers on the lines.

2, __ , __ , __ ,10, __ , __ , __ ,18, __

Skip count by 10 to find how many lollipops.

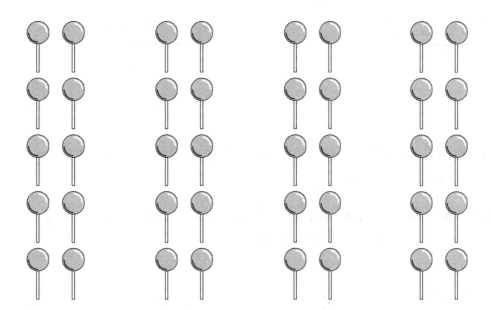

_____ lollipops

Build and write.

$$\begin{array}{r} 5 \\ +\ 5 \\ \hline \end{array}$$

$$\begin{array}{r} 1 \\ +\ 6 \\ \hline \end{array}$$

$$\begin{array}{r} 40 \\ +\ 10 \\ \hline \end{array}$$

$$\begin{array}{r} 10 \\ +\ 60 \\ \hline \end{array}$$

Read the word problem to the student if necessary. Use the pictures to help solve it.

Bob saw three stars. Bill saw three stars, too. How many stars did Bob and Bill see in all?

☆☆☆ + ☆☆☆ = _____ stars

Begin at the star by the number 10. Skip count by ten and connect the dots to finish the picture.

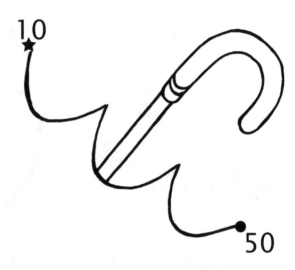

20 •

•
40

30

Begin at the star by the number 10. Skip count by ten and connect the dots to finish the picture.

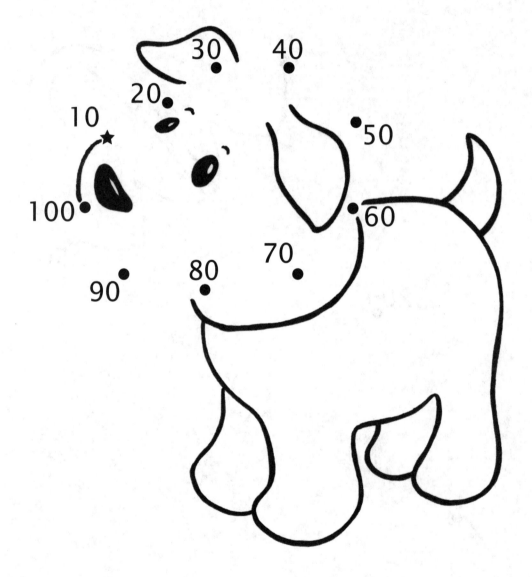

Build and add the hundreds. Say the problem as you write it.

<u>100</u> + <u>100</u> = _____

_____ + _____ = _____

Build and add the hundreds. Write and say the answers.

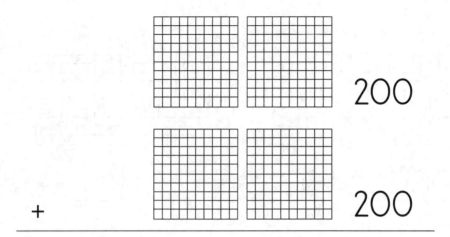

200

+   200
_____

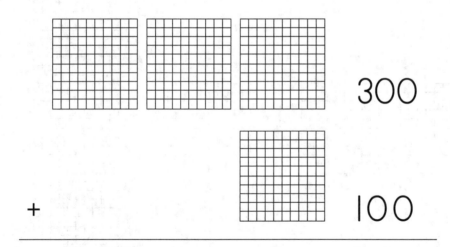

300

+   100
_____

Build and add the hundreds. Say the problem as you write it.

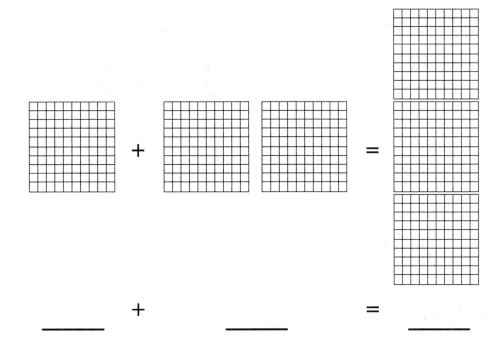

____ + ____ = ____

____ + ____ = ____

Build and add the hundreds. Write and say the answers.

```
    1 0 0              2 0 0
 + 3 0 0            + 2 0 0
 ─────────          ─────────
```

```
    1 0 0
 + 1 0 0
 ─────────
```

Read the word problem to the student if necessary. You may use the blocks to help you solve it.

Sam counted 100 red cars and 200 blue cars. How many cars did Sam count in all?

$100 + 200 =$ _____ cars

LESSON PRACTICE

Build and add the hundreds. Say the problem as you write it.

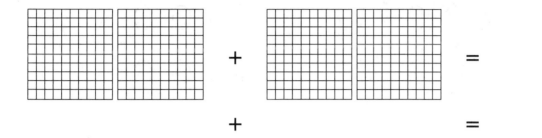

_____ + _____ = _____

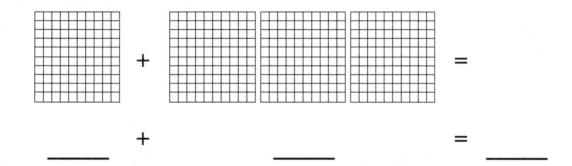

_____ + _____ = _____

Build and add the hundreds. Write and say the answers.

$$\begin{array}{r} 2\ 0\ 0 \\ +\ 1\ 0\ 0 \\ \hline \end{array} \qquad \begin{array}{r} 1\ 0\ 0 \\ +\ 1\ 0\ 0 \\ \hline \end{array}$$

$$\begin{array}{r} 3\ 0\ 0 \\ +\ 1\ 0\ 0 \\ \hline \end{array}$$

Read the word problem to the student if necessary. You may use the blocks to help you solve it.

200 boys and 200 girls went to school today. How many children went to school?

200 + 200 = _____ children

Build and add the hundreds. Write and say the answers.

$$\begin{array}{r} 1\ 0\ 0 \\ +\ 3\ 0\ 0 \\ \hline \end{array}$$

$$\begin{array}{r} 2\ 0\ 0 \\ +\ 2\ 0\ 0 \\ \hline \end{array}$$

$$\begin{array}{r} 2\ 0\ 0 \\ +\ 1\ 0\ 0 \\ \hline \end{array}$$

Skip count by 10 to find how many circles.

_____ circles

Skip count by 10. Write and say the missing numbers.

10, __ , __ , __ , __ , 60, __ , __ , 90, __

Build and write.

$$\begin{array}{r} 5 \\ +5 \\ \hline \end{array}$$
$$\begin{array}{r} 4 \\ +4 \\ \hline \end{array}$$

$$\begin{array}{r} 80 \\ +10 \\ \hline \end{array}$$
$$\begin{array}{r} 10 \\ +30 \\ \hline \end{array}$$

Read the word problem to the student if necessary. Use the pictures to help solve it.

Mom gave Dad two gifts. Jill gave Dad one gift. How many gifts did Dad get?

🎁 🎁 + 🎁 = _____ gifts

20E

Build and add the hundreds. Write and say the answers.

$$\begin{array}{r} 1\ 0\ 0 \\ +\ 2\ 0\ 0 \\ \hline \end{array}$$

$$\begin{array}{r} 3\ 0\ 0 \\ +\ 1\ 0\ 0 \\ \hline \end{array}$$

$$\begin{array}{r} 1\ 0\ 0 \\ +\ 1\ 0\ 0 \\ \hline \end{array}$$

Skip count by two to find how many birds.

_____ birds

Skip count by two. Write and say the missing numbers.

2, __ , __ , 8, __ , __ , __ , 16, __ , 20

Build and write.

$$10 + 50 = \underline{} \qquad\qquad 200 + 200 = \underline{}$$

$$\begin{array}{r} 7 \\ + \ 1 \\ \hline \end{array} \qquad\qquad\qquad \begin{array}{r} 1 \\ + \ 4 \\ \hline \end{array}$$

$$\begin{array}{r} 2\ 0 \\ + \ 2\ 0 \\ \hline \end{array} \qquad\qquad\qquad \begin{array}{r} 1\ 0 \\ + \ 1\ 0 \\ \hline \end{array}$$

Read the word problem to the student if necessary. You may use the blocks to help you solve it.

Luke ate 10 nuts before lunch. He ate 40 nuts after lunch. How many nuts did Luke eat in all?

$$10 + 40 = \underline{} \text{ nuts}$$

Build and add the hundreds. Write and say the answers.

$$\begin{array}{r} 2\ 0\ 0 \\ +\ 2\ 0\ 0 \\ \hline \end{array}$$
$$\begin{array}{r} 2\ 0\ 0 \\ +\ 1\ 0\ 0 \\ \hline \end{array}$$

$$\begin{array}{r} 3\ 0\ 0 \\ +\ 1\ 0\ 0 \\ \hline \end{array}$$

Skip count by two to find how many rectangles.

_____ rectangles

Skip count by two. Write and say the missing numbers.

__ , __ , 6, __ , __ , 12, __ , __ , 18, __

Build and write.

$$70 + 10 = \_\_\_ \qquad\qquad 100 + 300 = \_\_\_$$

$$\begin{array}{r} 3 \\ + 3 \\ \hline \end{array} \qquad\qquad \begin{array}{r} 1 \\ + 1 \\ \hline \end{array}$$

$$\begin{array}{r} 10 \\ + 20 \\ \hline \end{array} \qquad\qquad \begin{array}{r} 40 \\ + 40 \\ \hline \end{array}$$

....

Read the word problem to the student if necessary. You may use the blocks to help you solve it.

Jan has five fingers on her right hand. She has five fingers on her left hand. How many fingers does Jan have in all?

$$5 + 5 = \_\_\_\_\_ \text{ fingers}$$

Add.

If the answer is 200, color the space red.
If the answer is 400, color the space blue.
If the answer is 600, color the space purple.

100 + 100	200 + 200	100 + 100
200 + 200	300 + 300	200 + 200
100 + 100	200 + 200	100 + 100

What shapes do you see?

To the parent: You may wish to have the student try to count the squares. This is challenging because of overlapping shapes. The answer is at the bottom of next page.

Begin at the star by the number 10. Skip count by ten and connect the dots to finish the picture.

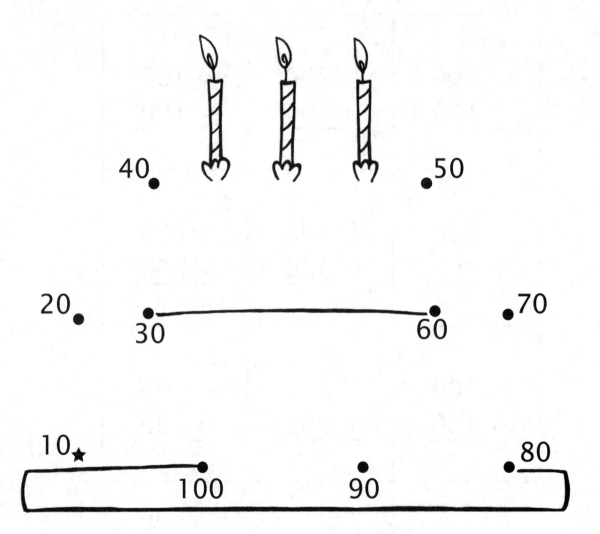

There are 14 squares on the previous page.

Use the blocks to solve for the unknown. Write the answer in the blank and say it.

_____ + 2 = 4

_____ + 3 = 6

_____ + 1 = 7

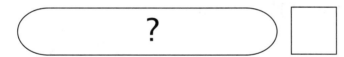

_____ + 5 = 10

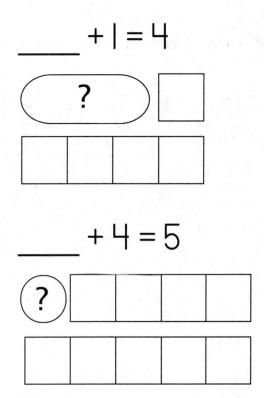

$$\underline{\quad} + 1 = 4$$

$$\underline{\quad} + 4 = 5$$

....

Read the word problem. Use the blocks to help you solve for the unknown number.

Nick wants three toy airplanes. He has two airplanes. How many more airplanes does Nick want?

$$\underline{\quad} + \text{✈ ✈} = \text{✈ ✈ ✈}$$

Use the blocks to solve for the unknown. Write the answer in the blank and say it.

_____ + 2 = 5

_____ + 3 = 7

_____ + 2 = 3

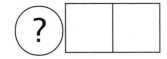

_____ + 1 = 9

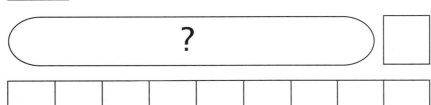

____ + 3 = 6

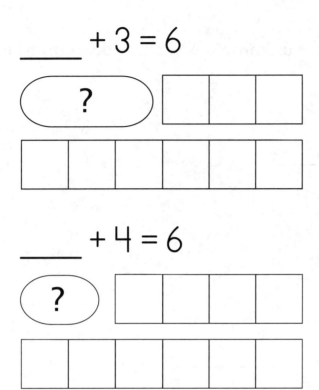

____ + 4 = 6

Read the word problem. Use the blocks to help you solve for the unknown number.

Jim and Dave ate four ice cream cones in all. Jim ate two cones. How many cones did Dave eat?

Use the blocks to solve for the unknown. Write the answer in the blank and say it.

_____ + 2 = 7

_____ + 3 = 4

_____ + 5 = 8

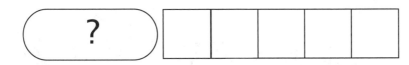

_____ + 1 = 6

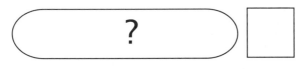

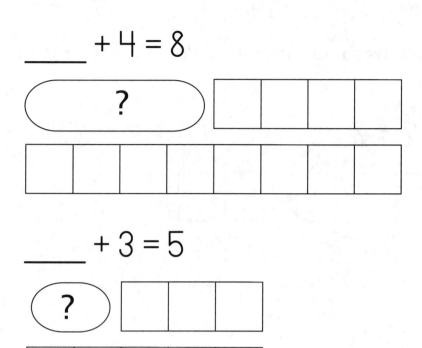

$$\underline{\quad} + 4 = 8$$

$$\underline{\quad} + 3 = 5$$

....

Read the word problem. Use the blocks to help you solve for the unknown number.

Sal drew six snowflakes. Mom helped her draw three of them. How many snowflakes did Sal draw by herself?

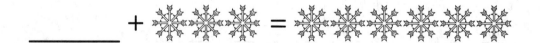

21D

Use the blocks to solve for the unknown. Write the answer in the blank and say it.

____ + 2 = 6

____ + 5 = 10

Build and write.

3 + 1 = ____

10 + 60 = ____

Build and write.

$$\begin{array}{r} 1 \\ +\ 8 \\ \hline \end{array}$$

$$\begin{array}{r} 2 \\ +\ 2 \\ \hline \end{array}$$

$$\begin{array}{r} 30 \\ +\ 10 \\ \hline \end{array}$$

$$\begin{array}{r} 100 \\ +100 \\ \hline \end{array}$$

Skip count by two to find how many triangles.

▽ ▽ ▽ ▽ ▽ ▽ ▽ ▽ ▽
▽ ▽ ▽ ▽ ▽ ▽ ▽ ▽ ▽

_____ triangles

Read the word problem. Use the blocks to help you solve for the unknown number.

Jed and Jack have two puppies. One puppy belongs to Jed. How many puppies belong to Jack?

_____ + = 🐶 🐶

Use the blocks to solve for the unknown. Write the answer in the blank and say it.

_____ + 5 = 6

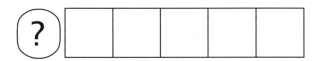

_____ + 2 = 8

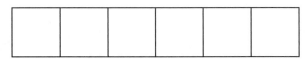

Build and write.

4 + 4 = _____

80 + 10 = _____

Build and write.

$$
\begin{array}{r}
6 \\
+\ 1 \\
\hline
\end{array}
\qquad
\begin{array}{r}
1 \\
+\ 5 \\
\hline
\end{array}
$$

$$
\begin{array}{r}
3\ 0 \\
+\ 3\ 0 \\
\hline
\end{array}
\qquad
\begin{array}{r}
2\ 0\ 0 \\
+\ 1\ 0\ 0 \\
\hline
\end{array}
$$

Skip count by 10 to find how many stars.

_____ stars

Use the blocks to solve for the unknown. Write the answer in the blank and say it.

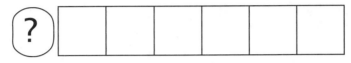

_____ + 6 = 7

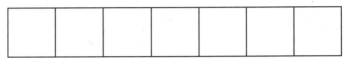

_____ + 3 = 6

Build and write.

| + | = _____

20 + 20 = _____

$$
\begin{array}{r}
7 \\
+\ 1 \\
\hline
\end{array}
\qquad
\begin{array}{r}
1 \\
+\ 2 \\
\hline
\end{array}
$$

$$
\begin{array}{r}
40 \\
+\ 40 \\
\hline
\end{array}
\qquad
\begin{array}{r}
300 \\
+\ 100 \\
\hline
\end{array}
$$

Skip count by two to find how many hearts.

_____ hearts

Read the word problem. Use the blocks to help you solve for the unknown number.

Jan needs 10 dollars. She has 5 dollars. How many more dollars does Jan need?

_____ + ☐ ☐ ☐ ☐ ☐ = ☐ ☐ ☐ ☐ ☐ ☐ ☐ ☐ ☐

Tom has two cookies.

Draw the cookies Tom needs to make four.

2 cookies + _____ cookies = 4 cookies

Sam has one cookie.

Draw the cookies Sam needs to make four.

1 cookie + _____ cookies = 4 cookies

Ava drew three triangles.

Draw the triangles Ava needs to make five.

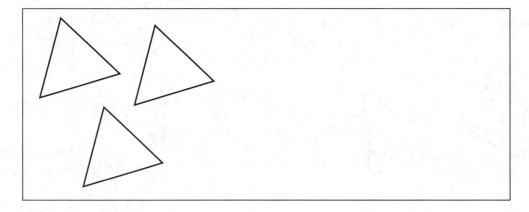

3 triangles + ___ triangles = 5 triangles

Pam drew four triangles.

Draw the triangles Pam needs to make five.

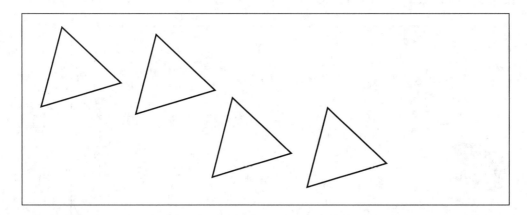

4 triangles + ___ triangle = 5 triangles

22A

Skip count by five and write the numbers on the lines. The first one has been done for you.

				__5__
				__10__
				__15__
				__20__
				__25__

				__5__
				____
				__15__
				____
				____
				__30__

Skip count by five and write the numbers on the lines. Then write the numbers in the spaces under the squares.

				___
				<u>10</u>
				___
				<u>20</u>
				___
				___
				<u>35</u>
				<u>40</u>

5, __ , 15, ___ , ___ , ___ , ___ , 40

22B

Skip count by five and write the numbers on the lines. Then write the numbers in the spaces under the squares.

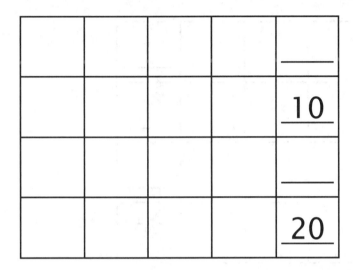

5, ____ , 15, ____

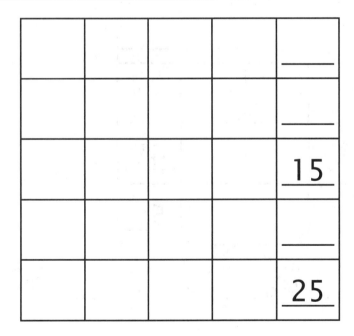

____ , 10, ____ , 20, ____

Skip count by five and write the numbers on the lines. Then write the numbers in the spaces under the squares.

				<u>5</u>
				<u> </u>
				<u> </u>
				<u>20</u>
				<u> </u>
				<u>30</u>
				<u> </u>
				<u> </u>
				<u>45</u>
				<u>50</u>

__ , 10, __ , __ , 25, __ , 35, 40, __ , 50

Skip count by five and write the numbers on the lines. Then write the numbers in the spaces under the squares.

				___
				___
				15
				___
				25
				___
				___
				40
				___
				___

5, __ , __ , __ , __ , 30, __ , __ , 45, 50

Skip count by five to find how many pencils.

_____ pencils

22D

Skip count by five and write the numbers on the lines. Then write the numbers in the spaces under the squares.

				____
				10
				____
				20
				____
				____
				35
				____
				____
				____

5, __ , 15, __ , __ , 30, __ , __ , 45, __

Skip count by five to find how many balloons.

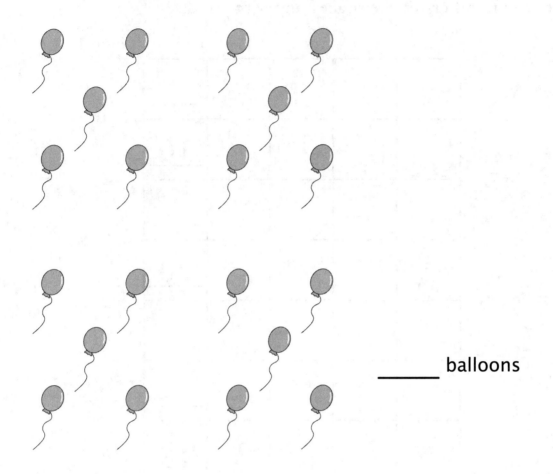

_____ balloons

Use the blocks to solve for the unknown. Write the answer in the blank and say it.

_____ $+ 2 = 9$

_____ $+ 4 = 5$

Skip count by five. Write the missing numbers on the lines.

__ , 10, __ , __ , 25, __ , 35, __ , __ , 50

....

Skip count by five to find how many flowers.

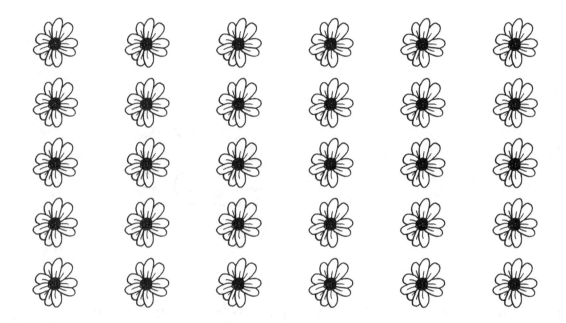

_____ flowers

....

Use the blocks to solve for the unknown. Write the answer in the blank and say it.

_____ + 4 = 7

_____ + 2 = 4

Build and write.

$$\begin{array}{r} 4 \\ +\ 1 \\ \hline \end{array} \qquad\qquad \begin{array}{r} 3 \\ +\ 3 \\ \hline \end{array}$$

$$\begin{array}{r} 1\ 0 \\ +\ 7\ 0 \\ \hline \end{array} \qquad\qquad \begin{array}{r} 2\ 0\ 0 \\ +\ 2\ 0\ 0 \\ \hline \end{array}$$

Read the word problem. Use the blocks to help you solve it.

Rob saw 10 cats. Jon saw 20 cats. How many cats did Rob and Jon see altogether?

$$10 + 20 = \underline{\hspace{1.5cm}} \text{ cats}$$

22F

Skip count by five. Write the missing numbers on the lines.

5, __ , __ , __ , __ , 30, __ , __ , 45, __

....

Skip count by five to find how many squares.

_____ squares

....

Use the blocks to solve for the unknown. Write the answer in the blank and say it.

____ + 3 = 9

____ + 1 = 3

Build and write.

$$\begin{array}{r} 5 \\ + \quad 5 \\ \hline \end{array} \qquad \begin{array}{r} 1 \\ + \quad 8 \\ \hline \end{array}$$

$$\begin{array}{r} 6\ 0 \\ + \quad 1\ 0 \\ \hline \end{array} \qquad \begin{array}{r} 1\ 0\ 0 \\ + 1\ 0\ 0 \\ \hline \end{array}$$

Read the word problem. Use the blocks to help you solve it.

One big fish and three little fish swim in the pond. How many fish swim in the pond?

$1 + 3 =$ _____ fish

Begin at the star by the number 5. Skip count by five and connect the dots to finish the picture.

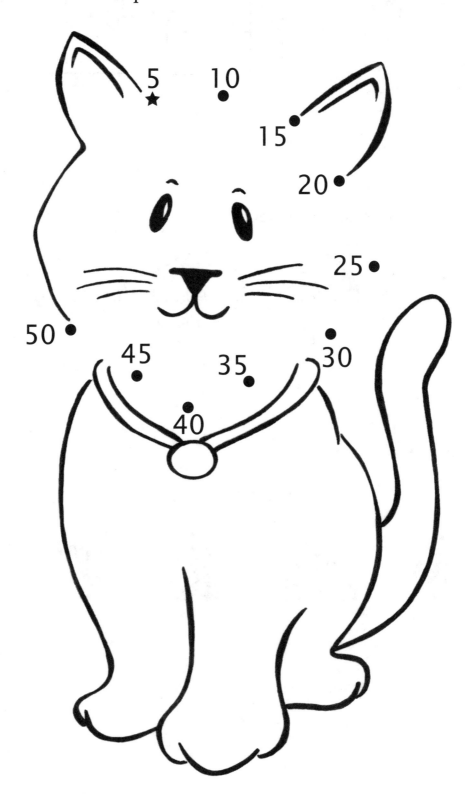

Look at the pictures carefully. Put an X on the one that is different.

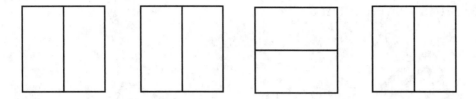

Put an X on the one that is different.

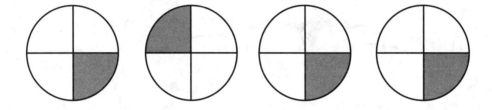

Put an X on the one that is different.

Write the numeral for each set of tally marks. The first one has been done for you.

𝍸 = 5

𝍸 𝍸 = ___

𝍸 | = ___

𝍸 𝍸 | = ___

𝍸 𝍸 𝍸 𝍸 = ___

𝍸 𝍸 𝍸 𝍸 𝍸 | = ___

Use tally marks to write the number. The first one has been done for you.

17 = ‖‖‖ ‖‖‖ ‖‖‖ ‖‖

10 = _____

6 = _____

15 = _____

Write the numeral for each set of tally marks.

||| = _____

||||| ||||| ||||| = _____

||||| = _____

||||| | = _____

||||| ||||| ||||| ||||| ||||| = _____

||||| ||||| ||||| ||||| ||||| ||||| ||||| = _____

Use tally marks to write the number.

20 = _____

16 = _____

11 = _____

7 = _____

23C

Write the numeral for each set of tally marks.

| | | | = ____

||| = ____

||| || = ___

||| ||| ||| ||| = ___

||| ||| ||| ||| ||| ||| = ___

||| ||| ||| ||| ||| ||| ||| ||| | = ___

Use tally marks to write the number.

12 = _____

21 = _____

25 = _____

30 = _____

Write the numeral.

ℸℸℸℸℸ ℸℸℸℸℸ ℸℸℸℸℸ I = _____

Use tally marks to write the number.

10 = _____

Skip count by five. Write the missing numbers on the lines.

5, __ , 15, __ , __ , 30, __ , __ , __ , 50

Use the blocks to solve for the unknown. Write the answer in the blank and say it.

_____ + 7 = 8

_____ + 3 = 5

Build and write.

```
    6            4
+   1        +   4
_____      _____
```

```
   5 0         1 0 0
+  1 0       + 2 0 0
_____     _____
```

Write the numeral.

$\cancel{||||}$ $\cancel{||||}$ $\cancel{||||}$ $\cancel{||||}$ $\cancel{||||}$ $\cancel{||||}$ $\cancel{||||}$ $\cancel{||||}$ $\cancel{||||}$ = ___

Use tally marks to write the number.

26 = _____

Skip count by two. Write the missing numbers on the lines.

2, __, __, 8, __, __, __, 16, __, __

Use the blocks to solve for the unknown. Write the answer in the blank and say it.

___ + 4 = 9

___ + 5 = 10

Build and write.

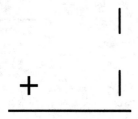

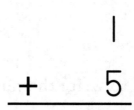

$$\begin{array}{r} 1\,0 \\ +\ 3\,0 \\ \hline \end{array}$$

$$\begin{array}{r} 2\,0\,0 \\ +\,2\,0\,0 \\ \hline \end{array}$$

Write the numeral.

IIII IIII IIII IIII IIII IIII I = _____

Use tally marks to write the number.

16 = _____

Skip count by ten. Write the missing numbers on the lines.

10, __, __, 40, __, __, __, __, __, 100

Use the blocks to solve for the unknown. Write the answer in the blank and say it.

____ + 1 = 5

____ + 6 = 8

Build and write.

$$\begin{array}{r} 1 \\ +\ 2 \\ \hline \end{array} \qquad\qquad \begin{array}{r} 2 \\ +\ 2 \\ \hline \end{array}$$

$$\begin{array}{r} 10 \\ +\ 40 \\ \hline \end{array} \qquad\qquad \begin{array}{r} 300 \\ +100 \\ \hline \end{array}$$

23G

Match the tally marks with the pictures.

Count the pictures. Use tally marks to write your answers.

_____

_____

_____

The diagrams are smaller than actual size. Have the student count to select the correct unit bars to complete the problems. Build, say, write, and color.

□ + □□□□□□□□ = □□□□□□□□□

_____ + _____ = _____

□□ + □□□□□□□ = □□□□□□□□□

_____ + _____ = _____

□□ + □□□□□□□□ = □□□□□□□□□□

_____ + _____ = _____

□□□□ + □□□□□□ = □□□□□□□□□□

_____ + _____ = _____

☐☐☐☐☐ + ☐☐☐☐☐ = ☐☐☐☐☐☐☐☐☐☐

_____ + _____ = _____

Build and write.

```
    4              1
+   6          +   9
_____        _____
```

```
    3              2
+   7          +   8
_____        _____
```

The diagrams are smaller than actual size. Have the student count to select the correct unit bars to complete the problems. Build, say, write, and color.

☐☐☐☐☐☐ + ☐☐☐ = ☐☐☐☐☐☐☐☐☐

_____ + _____ = _____

☐☐☐☐☐ + ☐☐☐☐☐ = ☐☐☐☐☐☐☐☐

_____ + _____ = _____

☐☐☐☐☐☐☐☐ + ☐☐ = ☐☐☐☐☐☐☐☐☐☐

_____ + _____ = _____

☐☐☐☐☐☐☐☐☐ + ☐ = ☐☐☐☐☐☐☐☐☐☐

_____ + _____ = _____

☐☐☐☐☐☐☐ + ☐☐☐ = ☐☐☐☐☐☐☐☐☐☐

_____ + _____ = _____

Build and write.

$$\begin{array}{r} 5 \\ + \ 5 \\ \hline \end{array} \qquad \begin{array}{r} 2 \\ + \ 8 \\ \hline \end{array}$$

$$\begin{array}{r} 7 \\ + \ 3 \\ \hline \end{array} \qquad \begin{array}{r} 9 \\ + \ 1 \\ \hline \end{array}$$

Use the blocks to solve for the unknown. Write the answer in the blank and say it.

____ + 5 = 10

____ + 4 = 10

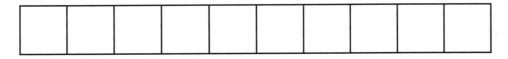

____ + 1 = 10

____ + 8 = 10

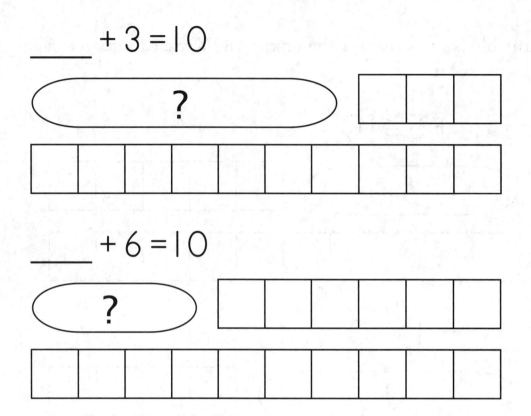

___ + 3 = 10

___ + 6 = 10

Build and write.

$$\begin{array}{r} 8 \\ +\ \ 2 \\ \hline \end{array}$$
$$\begin{array}{r} 3 \\ +\ \ 7 \\ \hline \end{array}$$

$$\begin{array}{r} 1 \\ +\ \ 9 \\ \hline \end{array}$$
$$\begin{array}{r} 5 \\ +\ \ 5 \\ \hline \end{array}$$

Use the blocks to solve for the unknown. Write the answer in the blank and say it.

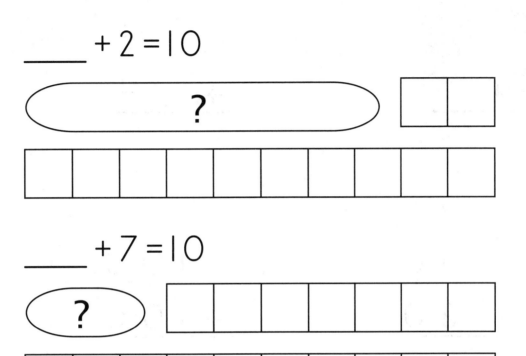

$\underline{} + 2 = 10$

$\underline{} + 7 = 10$

Write the numeral.

$\overline{|||}\ \overline{|||}\ \overline{|||}\ \overline{|||}\ \overline{|||} = \underline{}$

Skip count by two to find how many gifts.

$\underline{}$ gifts

Build and write.

$$
\begin{array}{r}
7 \\
+\ 1 \\
\hline
\end{array}
\qquad\qquad
\begin{array}{r}
6 \\
+\ 4 \\
\hline
\end{array}
$$

$$
\begin{array}{r}
3\,0 \\
+\ 3\,0 \\
\hline
\end{array}
\qquad\qquad
\begin{array}{r}
1\,0\,0 \\
+\,1\,0\,0 \\
\hline
\end{array}
$$

Read the word problem. Use the blocks to help you solve it.

Pam had three dimes. She found three more dimes. How many dimes does Pam have now?

$3 + 3 =$ _____ dimes

Use the blocks to solve for the unknown. Write the answer in the blank and say it.

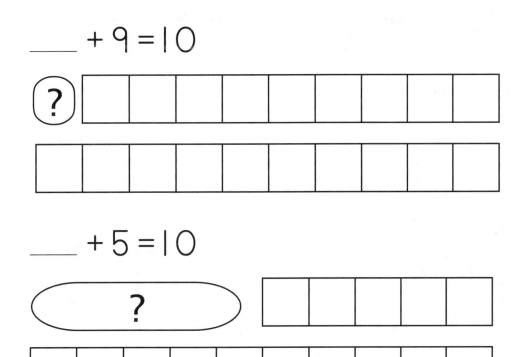

___ + 9 = 10

___ + 5 = 10

Skip count by five to find how many stars.

_____ stars

Use tally marks to write the number.

31 = _____

Build and write.

$$\begin{array}{r} 2 \\ + \quad 8 \\ \hline \end{array}$$

$$\begin{array}{r} 7 \\ + \quad 3 \\ \hline \end{array}$$

$$\begin{array}{r} 8\;0 \\ + \;1\;0 \\ \hline \end{array}$$

$$\begin{array}{r} 2\;0\;0 \\ + \;1\;0\;0 \\ \hline \end{array}$$

Read the word problem. Use the blocks to help you solve it.

Twenty trees have green leaves. Twenty trees have brown leaves. How many trees are there in all?

20 + 20 = _____ trees

Use the blocks to solve for the unknown. Write the answer in the blank and say it.

____ + 6 = 10

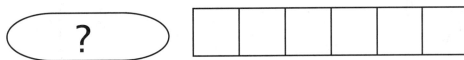

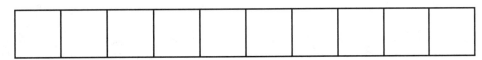

____ + 3 = 10

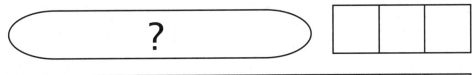

Skip count by 10 to find how many circles.

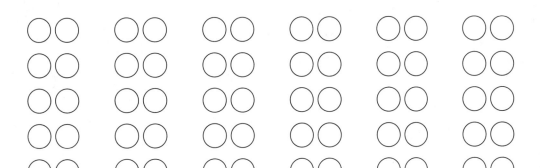

____ circles

Write the numeral.

𝍸𝍸 𝍸𝍸 𝍸𝍸 𝍸𝍸 𝍸𝍸 𝍸𝍸 𝍸𝍸 𝍸𝍸 𝍸𝍸 | = ___

Build and write.

```
    5              9
+   5          +   1
_____        _____
```

```
   40            200
+  40          + 200
_____        _____
```

Draw the circles you need to make 10.

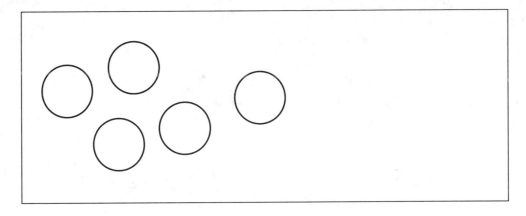

5 + _____ = 10

Draw the squares you need to make 10.

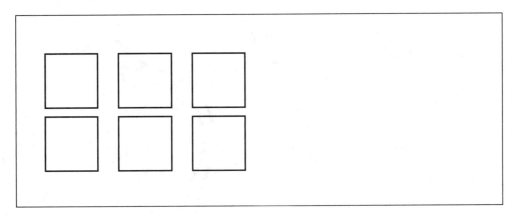

6 + _____ = 10

Begin at the star by the number 5. Skip count by five and connect the dots to finish the picture.

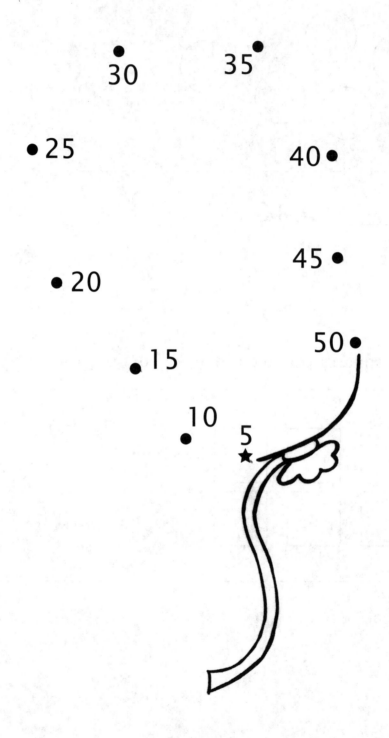

Find the area by skip counting. Write the numbers on the lines if you need to. Write your answer in the last box.

Find the area by skip counting. Write the numbers on the lines if you need to. Write your answer in the last box.

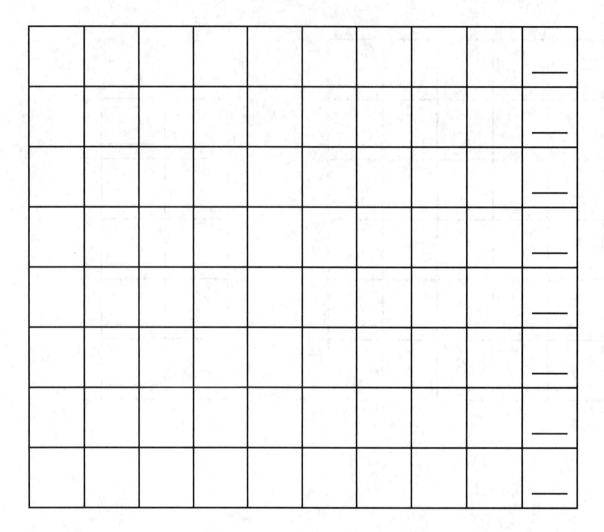

25B

Find the area by skip counting. Write the numbers on the lines if you need to. Write your answer in the last box.

Find the area by skip counting. Write the numbers on the lines if you need to. Write your answer in the last box.

LESSON PRACTICE

Find the area by skip counting. Write the numbers on the lines if you need to. Write your answer in the last box.

Find the area by skip counting. Write the numbers on the lines if you need to. Write your answer in the last box.

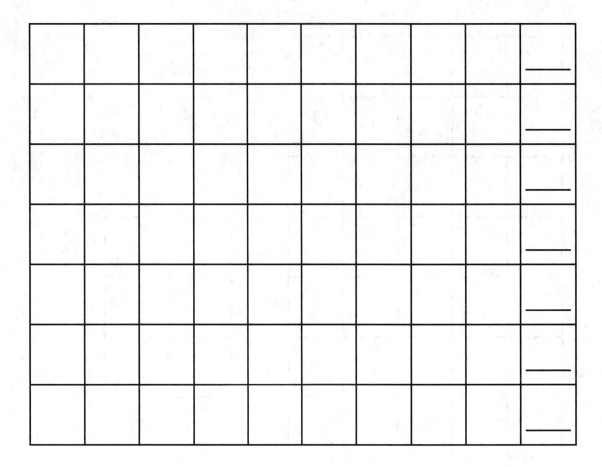

Find the area by skip counting. Write the numbers on the lines if you need to. Write your answer in the last box.

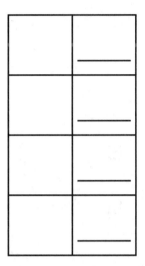 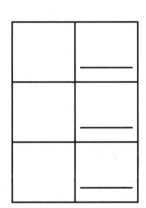

Write the numeral.

||||| ||||| ||||| = _____

- -

Build and write.

$$\begin{array}{r} 2 \\ + 8 \\ \hline \end{array}$$
$$\begin{array}{r} 4 \\ + 4 \\ \hline \end{array}$$

$$\begin{array}{r} 8 \\ + 1 \\ \hline \end{array}$$
$$\begin{array}{r} 70 \\ + 10 \\ \hline \end{array}$$

- -

Read the word problem. Use the blocks to help you solve it.

Four birds sang to me. One more bird came to sing. How many birds are singing now?

$4 + 1 =$ _____ birds

Find the area by skip counting. Write the numbers on the lines if you need to. Write your answer in the last box.

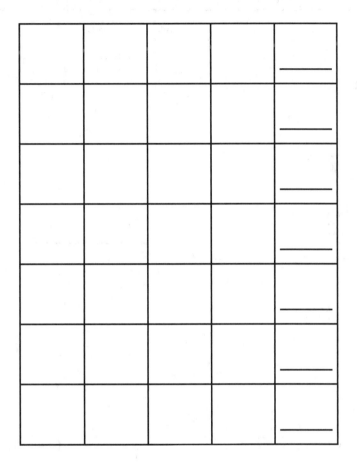

Use tally marks to write the number.

16 = _____

Build and write.

$$\begin{array}{r} 7 \\ +3 \\ \hline \end{array}$$

$$\begin{array}{r} 2 \\ +2 \\ \hline \end{array}$$

$$\begin{array}{r} 6 \\ +4 \\ \hline \end{array}$$

$$\begin{array}{r} 10 \\ +60 \\ \hline \end{array}$$

Read the word problem. Use the blocks to help you solve it.

Tom counted 300 bugs. Sam counted 100 bugs. How many bugs did Tom and Sam count in all?

300 + 100 = _____ bugs

25F

Find the area by skip counting. Write the numbers on the lines if you need to. Write your answer in the last box.

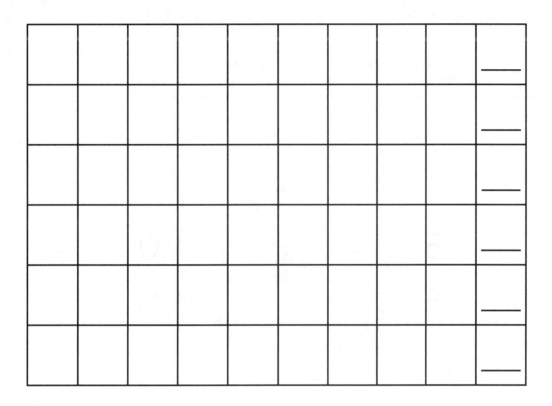

Write the numeral.

|||| |||| |||| |||| |||| |||| |||| |||| |||| | = _____

Build and write.

$$\begin{array}{r} 1 \\ +\ 9 \\ \hline \end{array}$$

$$\begin{array}{r} 1 \\ +\ 6 \\ \hline \end{array}$$

$$\begin{array}{r} 5 \\ +\ 5 \\ \hline \end{array}$$

$$\begin{array}{r} 10 \\ +\ 10 \\ \hline \end{array}$$

Read the word problem. Use the blocks to help you solve it.

Deb took 10 big steps. She took 50 little steps. How many steps did Deb take in all?

$10 + 50 =$ _____ steps

Draw the circles you need to make 10.

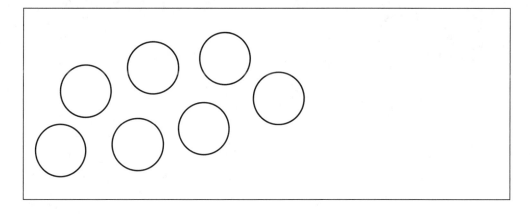

7 + _____ = 10

Draw the squares you need to make 10.

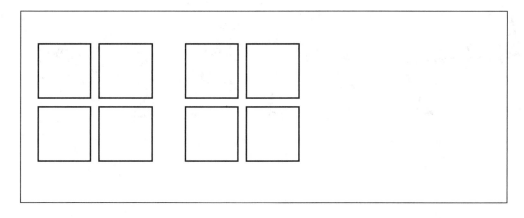

8 + _____ 10

Start at 5. Skip count by five and connect the dots to finish the picture. Help the student count by fives to 60.

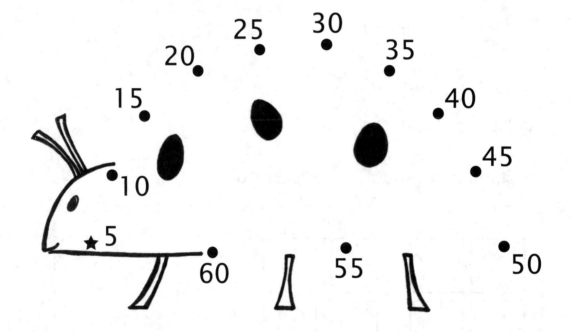

Count the minutes. There is a removable clock template near the end of this book.

———————

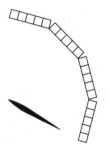

———————

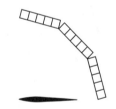

———————

———————

Write how many minutes are shown by each clock.

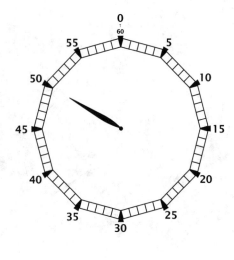

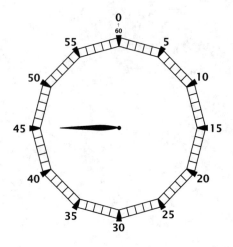

_____   _____

Count the minutes.

_____   _____

. . . .

Write how many minutes are shown by each clock.

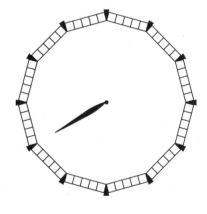

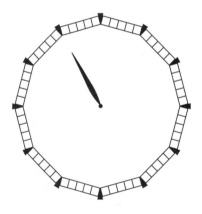

_____   _____

Write how many minutes are shown by each clock.

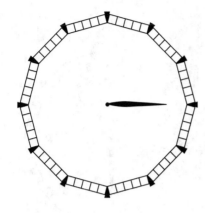

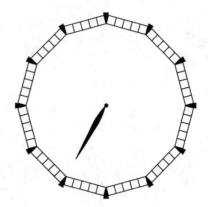

_____   _____

LESSON PRACTICE

Write how many minutes are shown by each clock.

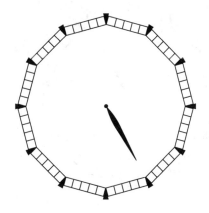

_____

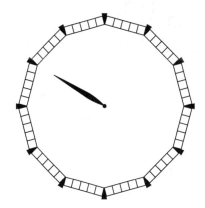

_____

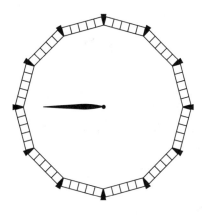

_____

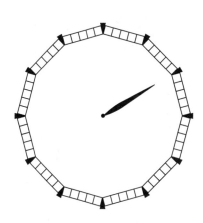

_____

Write how many minutes are shown by each clock.

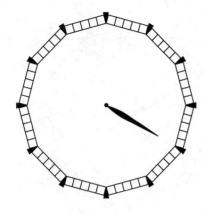 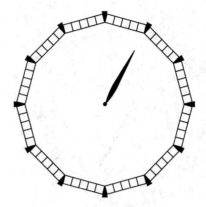

_____   _____

Write how many minutes are shown by each clock.

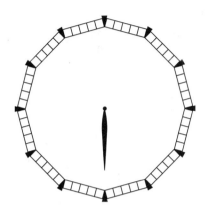

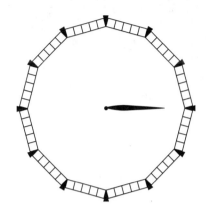

_____

_____

Find the area of each rectangle by skip counting. Write your answer in the last box.

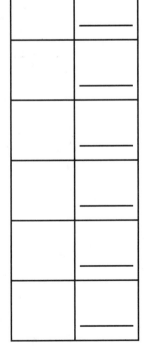

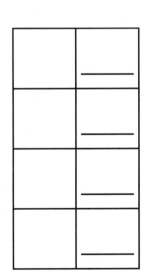

Write the numeral.

𝗍𝗁𝗅 𝗍𝗁𝗅 𝗍𝗁𝗅 | = _____

Build and write.

$$\begin{array}{r} 3 \\ + 3 \\ \hline \end{array}$$

$$\begin{array}{r} 4 \\ + 6 \\ \hline \end{array}$$

$$\begin{array}{r} 8 \\ + 1 \\ \hline \end{array}$$

$$\begin{array}{r} 20 \\ + 10 \\ \hline \end{array}$$

Write how many minutes are shown by each clock.

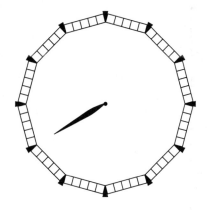 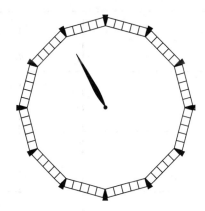

_____   _____

Find the area of each rectangle by skip counting. Write your answer in the last box.

				_____
				_____
				_____

Find the area of each rectangle by skip counting. Write your answer in the last box.

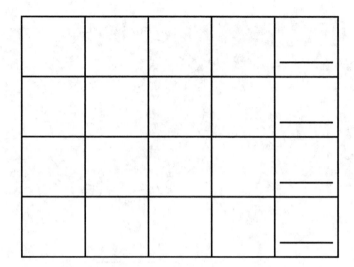

Look at the tally marks. Write the numeral.

|| = _____

Build and write.

$$\begin{array}{r} 6 \\ +\ 1 \\ \hline \end{array}$$

$$\begin{array}{r} 5 \\ +\ 5 \\ \hline \end{array}$$

$$\begin{array}{r} 1 \\ +\ 7 \\ \hline \end{array}$$

$$\begin{array}{r} 1\,0\,0 \\ +1\,0\,0 \\ \hline \end{array}$$

Write how many minutes are shown by each clock.

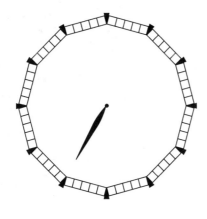 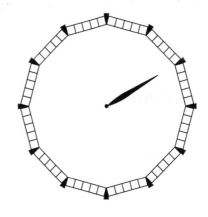

_____   _____

Find the area of the rectangle by skip counting. Write your
answer in the last box.

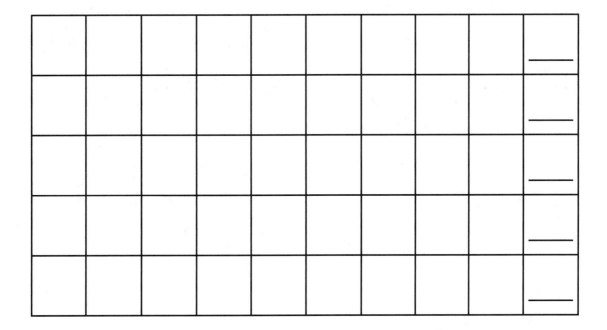

Write the numeral.

𝗜𝗡𝗜 𝗜𝗡𝗜 𝗜𝗡𝗜 𝗜𝗡𝗜 𝗜 = _____

Build and write.

```
      7                4
  +   3            +   4
  _____         _____


      1                1 0
  +   9            + 3 0
  _____         _____
```

Start at 5. Skip count by five and connect the dots to finish the picture. Help the student count by fives to 60.

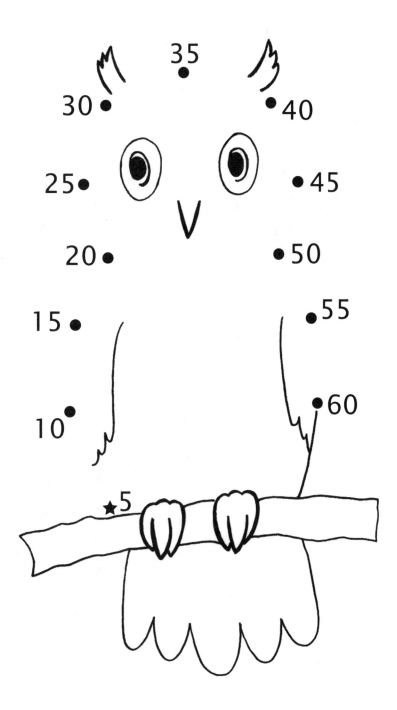

Start at 5. Skip count by five and connect the dots to finish the picture. Help the student count by fives to 70.

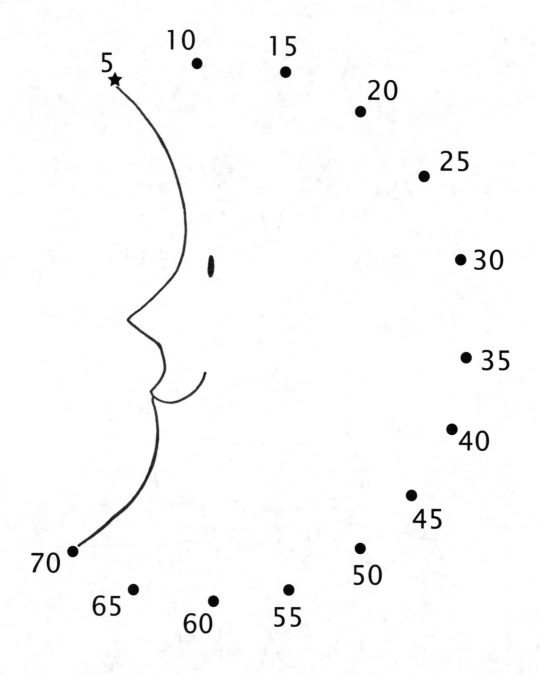

LESSON PRACTICE

What is the hour?

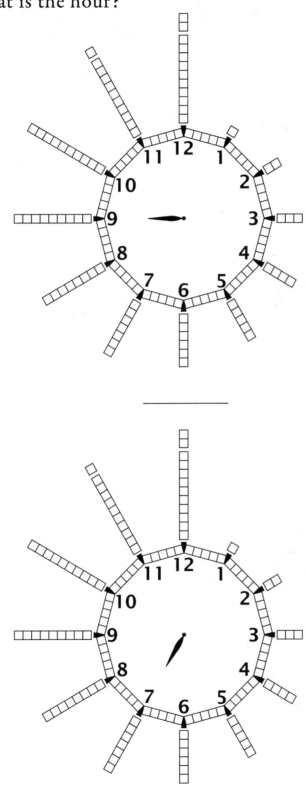

_____

_____

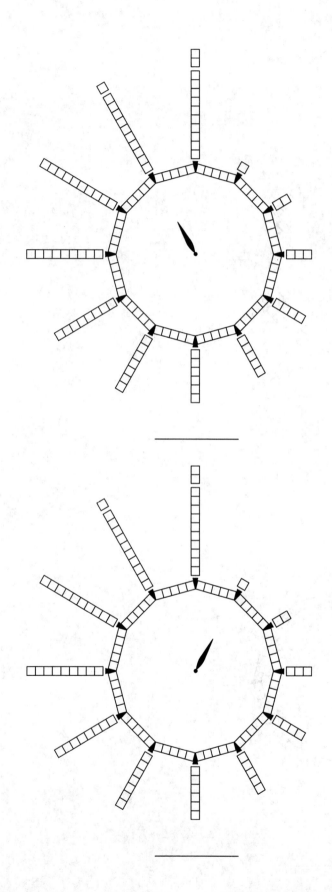

_____

_____

LESSON PRACTICE

What is the hour?

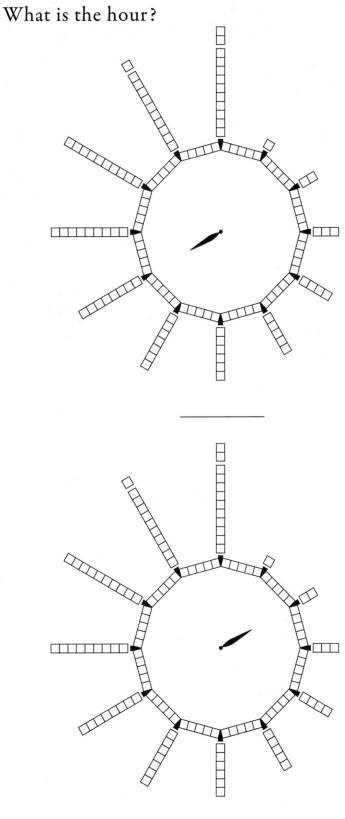

_____

_____

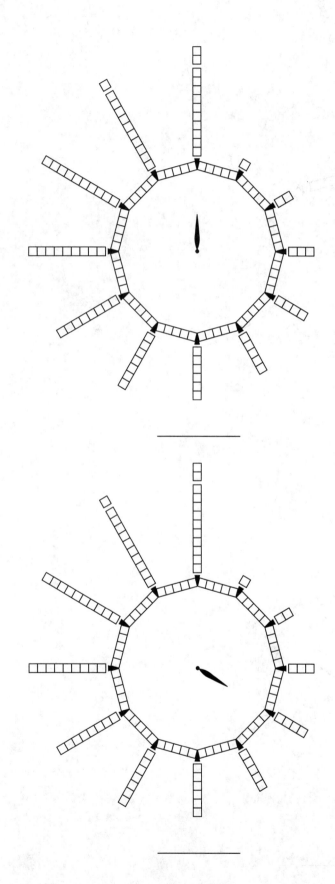

_____

_____

LESSON PRACTICE

What is the hour?

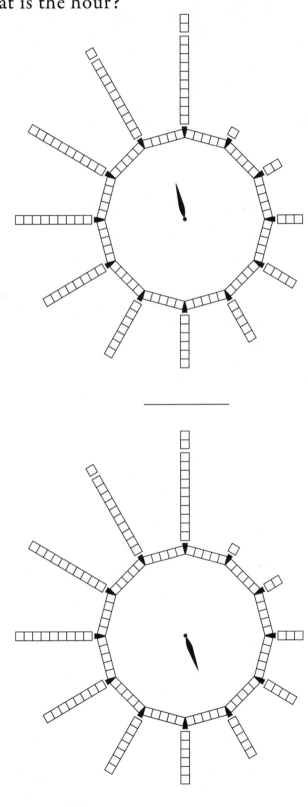

_____

_____

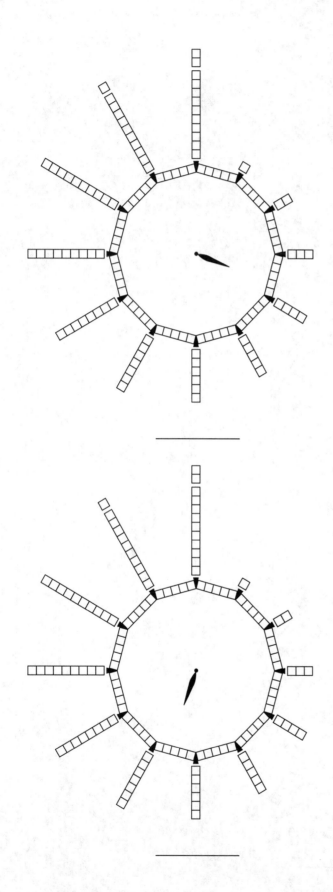

_____

_____

What is the hour?

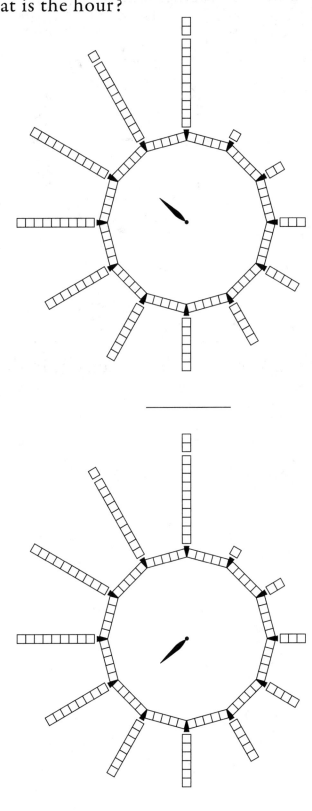

_____

_____

Write how many minutes are shown by each clock.

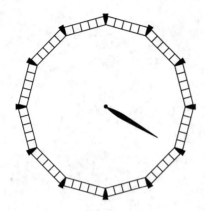

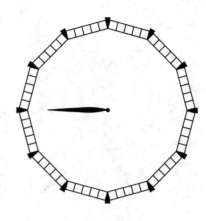

_____

_____

....

Use the blocks to solve for the unknown. Write the answer in the blank and say it.

___ + 5 = 7

___ + 1 = 6

What is the hour?

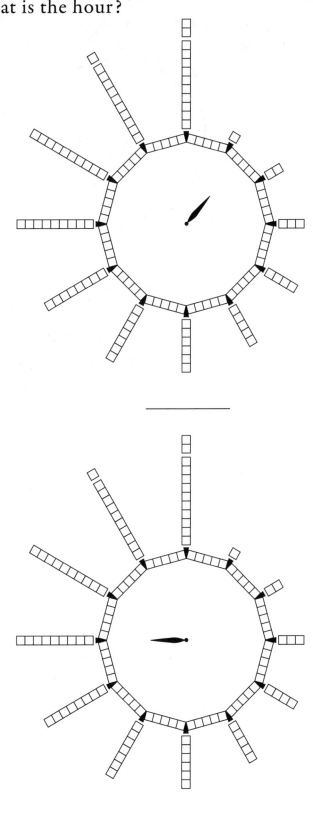

_____

_____

Write how many minutes are shown by each clock.

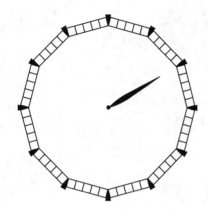 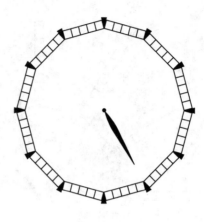

_____   _____

Use the blocks to solve for the unknown. Write the answer in the blank and say it.

$$\text{\_\_\_} + 3 = 8$$

$$\text{\_\_\_} + 2 = 3$$

What is the hour?

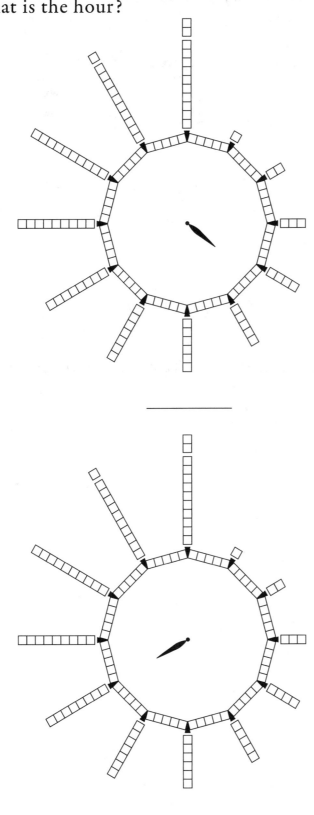

Write how many minutes are shown by each clock.

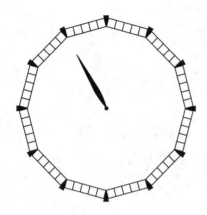 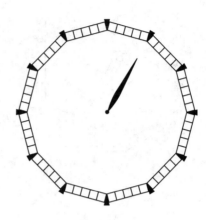

_____   _____

Use the blocks to solve for the unknown. Write the answer in the blank and say it.

____ + 5 = 9

____ + 6 = 7

Start at 5. Skip count by five. Connect the dots to finish the picture. Help the student count by fives to 100.

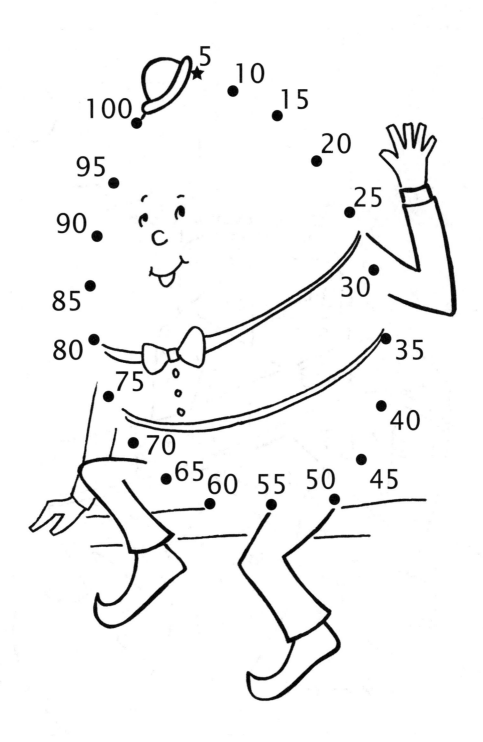

Add.

If the answer is 2, color the space yellow.
If the answer is 4, color the space blue.
If the answer is 6, color the space red.

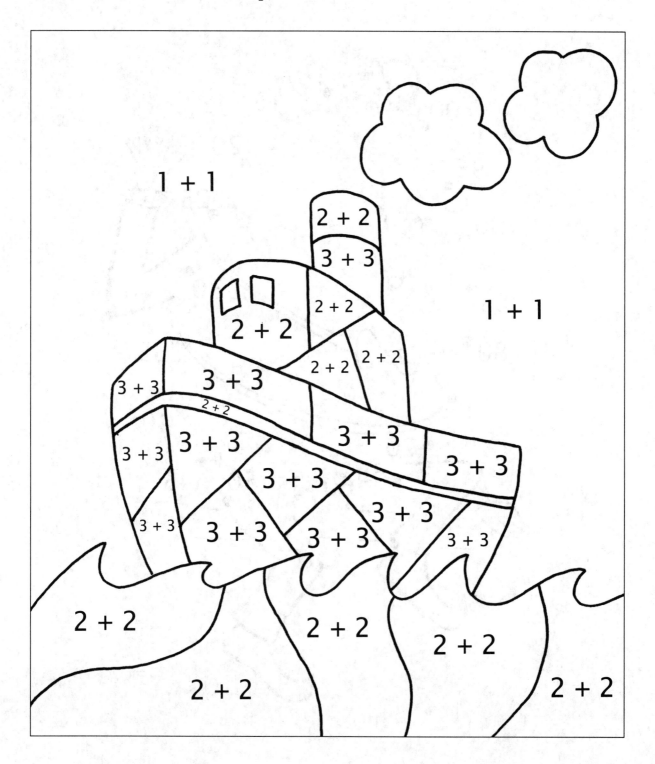

LESSON PRACTICE

Give the time with hours and minutes.

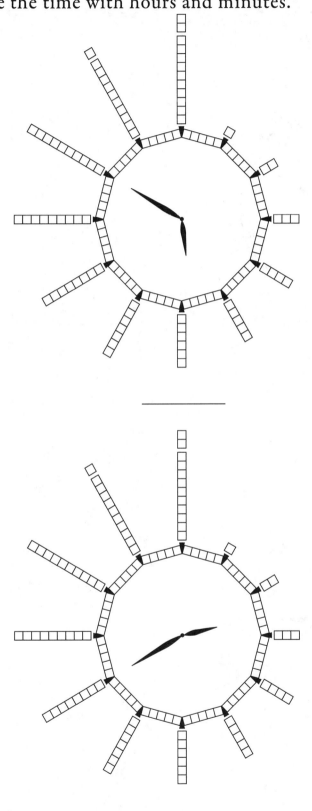

_____

_____

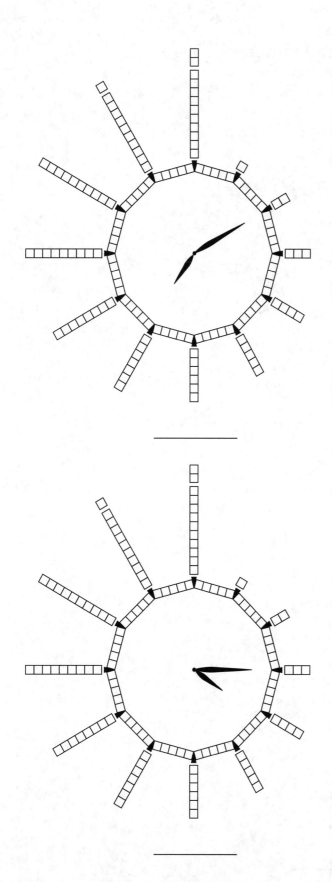

_____

_____

Give the time with hours and minutes.

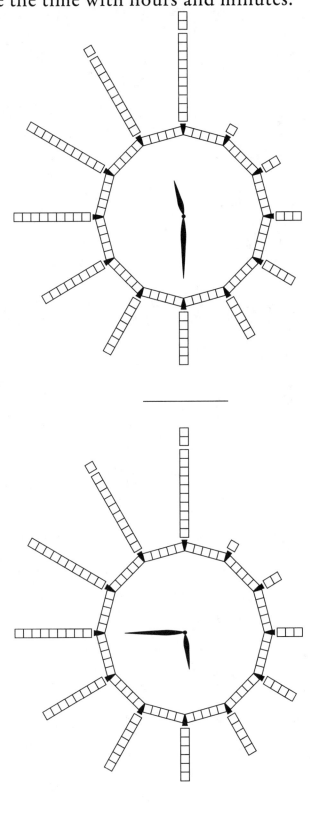

_____

_____

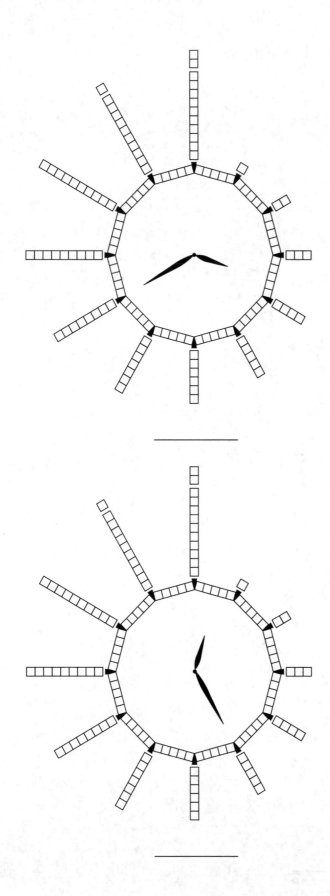

_____

_____

LESSON PRACTICE

Give the time with hours and minutes.

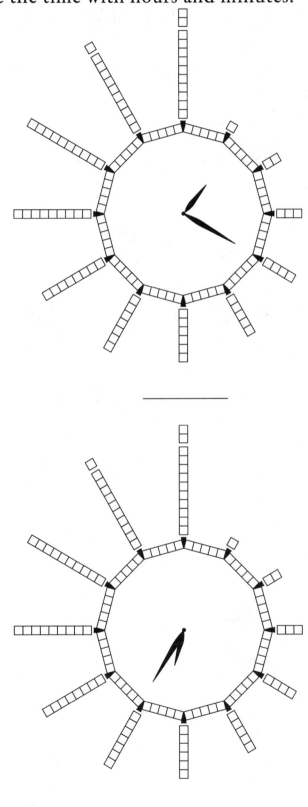

_____

_____

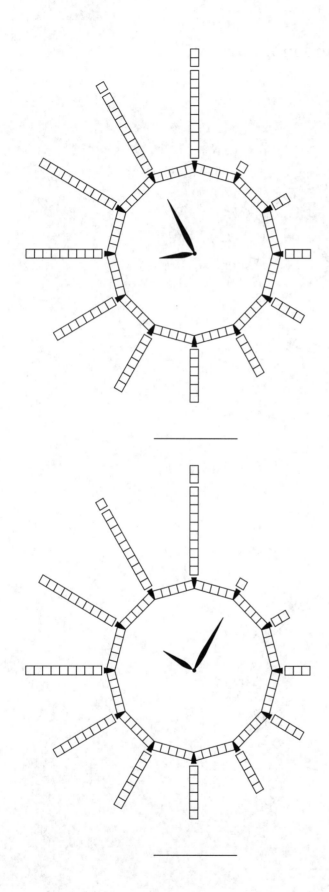

_____

_____

Give the time with hours and minutes.

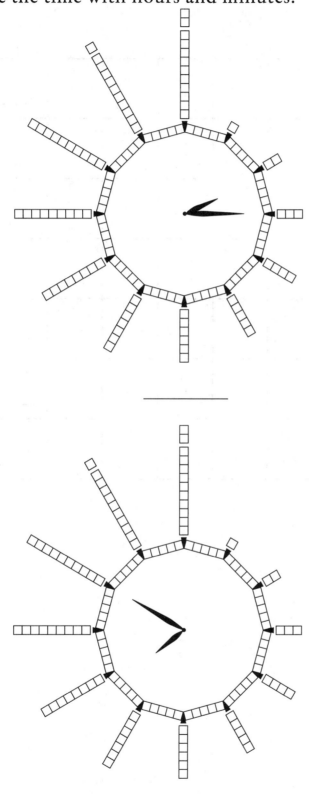

_____

_____

Find the area of the rectangle by skip counting. Write your answer in the last box.

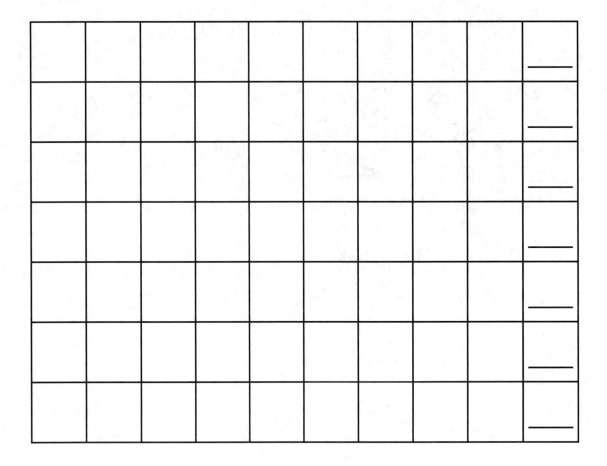

Give the time with hours and minutes.

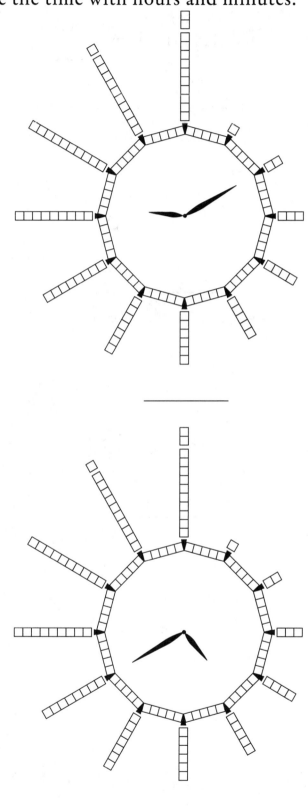

_____

_____

Write the numeral.

卌 卌 卌 卌 卌 卌 卌 卌 = _____

Build and write.

$$
\begin{array}{r}
1 \\
+\ 8 \\
\hline
\end{array}
\qquad
\begin{array}{r}
3 \\
+\ 3 \\
\hline
\end{array}
$$

$$
\begin{array}{r}
8 \\
+\ 2 \\
\hline
\end{array}
\qquad
\begin{array}{r}
50 \\
+\ 10 \\
\hline
\end{array}
$$

Give the time with hours and minutes.

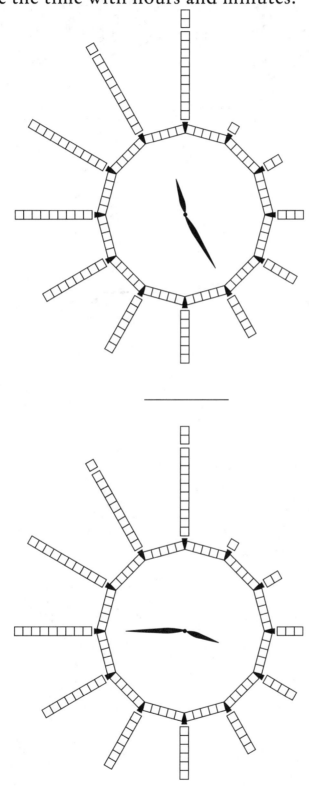

_____

_____

Write the numeral.

𝍸𝍸𝍸 𝍸𝍸𝍸 𝍸𝍸𝍸 𝍸𝍸𝍸 𝍸𝍸𝍸 𝍸𝍸𝍸 𝍸𝍸𝍸 𝍸𝍸𝍸 𝍸𝍸𝍸 𝍸𝍸𝍸 = _____

Build and write.

$$
\begin{array}{r} 3 \\ + \ 1 \\ \hline \end{array}
\qquad\qquad
\begin{array}{r} 6 \\ + \ 4 \\ \hline \end{array}
$$

$$
\begin{array}{r} 1 \\ + \ 5 \\ \hline \end{array}
\qquad\qquad
\begin{array}{r} 20 \\ + \ 20 \\ \hline \end{array}
$$

Add.

If the answer is 4, color the space light blue.
If the answer is 6, color the space brown.
If the answer is 8, color the space yellow.

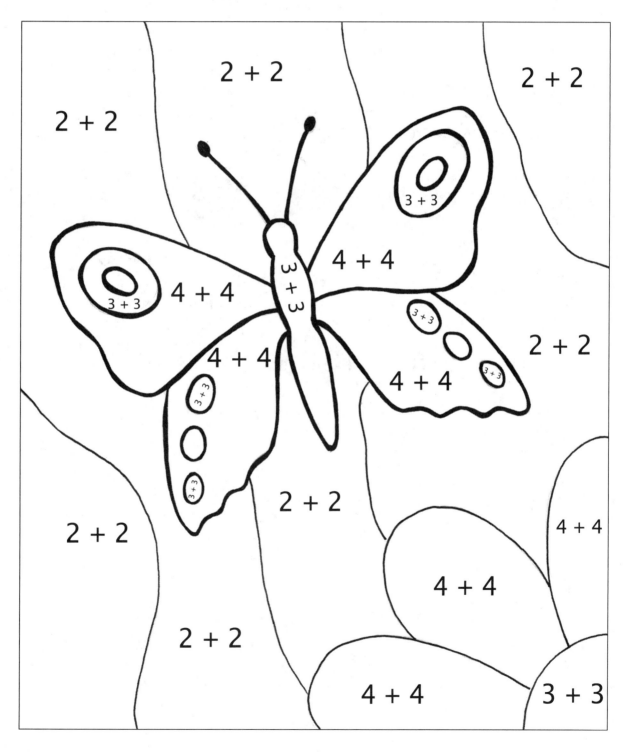

See if you can start at 20 and connect the dots by counting
backward. If this is too hard, start at 1 and connect the dots. Then
start at 20 and trace the line backward. Say the numbers as you go.

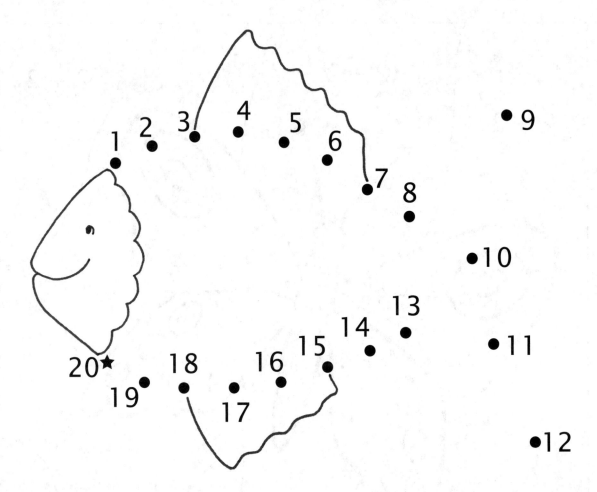

Build each problem and write the numbers in the circles. The first one has been done for you.

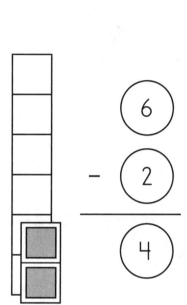

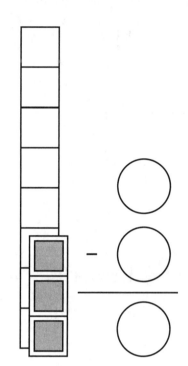

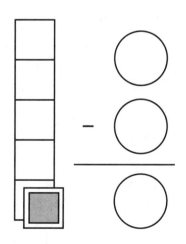

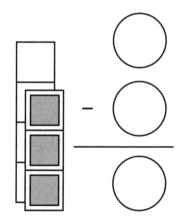

Build each problem. Say and write the answer.

$$\begin{array}{r} 4 \\ -\ 1 \\ \hline \end{array} \qquad \begin{array}{r} 5 \\ -\ 3 \\ \hline \end{array}$$

29B

Build each problem and write the numbers in the circles.

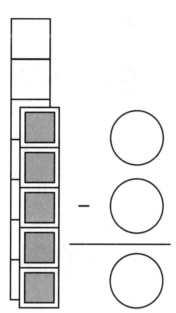

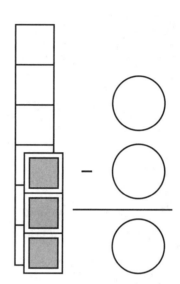

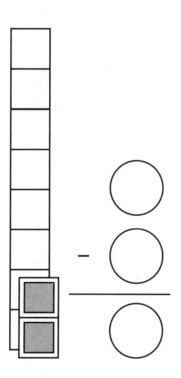

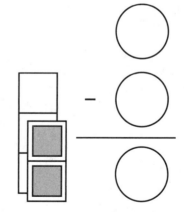

Build each problem. Say and write the answer.

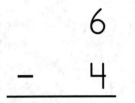

$$
\begin{array}{r}
6 \\
-\ 4 \\
\hline
\end{array}
\qquad
\begin{array}{r}
7 \\
-\ 6 \\
\hline
\end{array}
$$

29C

Build each problem and write the numbers in the circles.

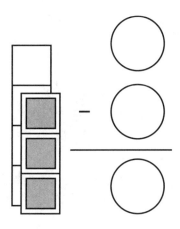

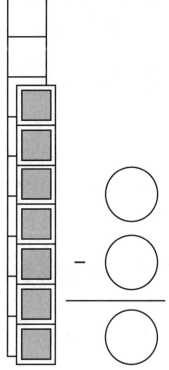

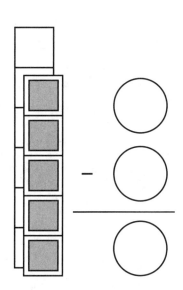

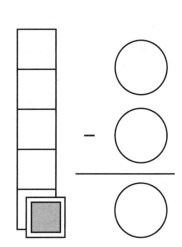

Build each problem. Say and write the answer.

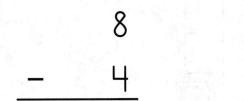

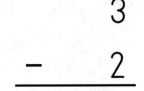

$$\begin{array}{r} 8 \\ -\ 4 \\ \hline \end{array} \qquad \begin{array}{r} 3 \\ -\ 2 \\ \hline \end{array}$$

Build each problem. Say and write the answer.

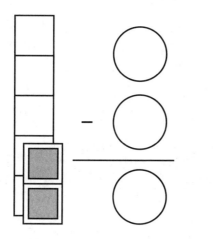 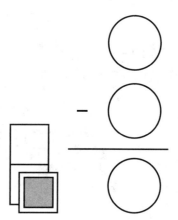

$$
\begin{array}{r}
7 \\
-\ 3 \\
\hline
\end{array}
\qquad
\begin{array}{r}
6 \\
-\ 1 \\
\hline
\end{array}
$$

$$
\begin{array}{r}
8 \\
-\ 6 \\
\hline
\end{array}
$$

Give the time with hours and minutes.

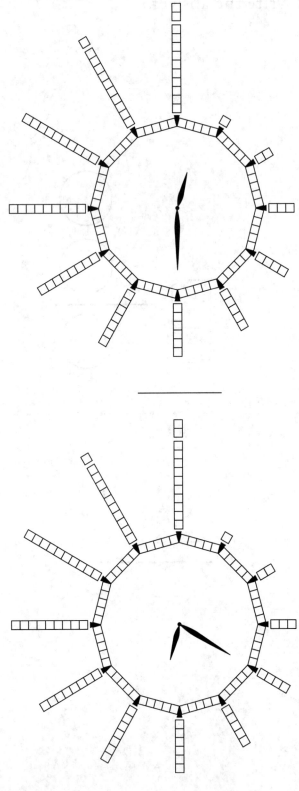

_____

_____

Build each problem. Say and write the answer.

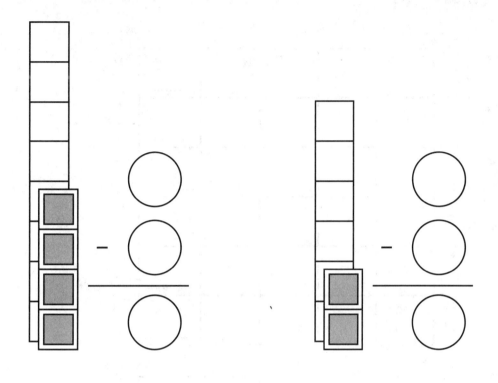

$$\begin{array}{r} 9 \\ -\ 2 \\ \hline \end{array}$$

$$\begin{array}{r} 6 \\ -\ 5 \\ \hline \end{array}$$

$$\begin{array}{r} 7 \\ -\ 4 \\ \hline \end{array}$$

Find the area of each rectangle by skip counting. Write your
answer in the last box.

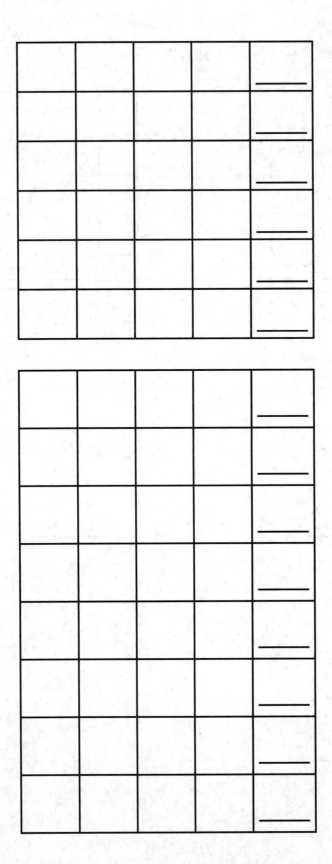

Build each problem. Say and write the answer.

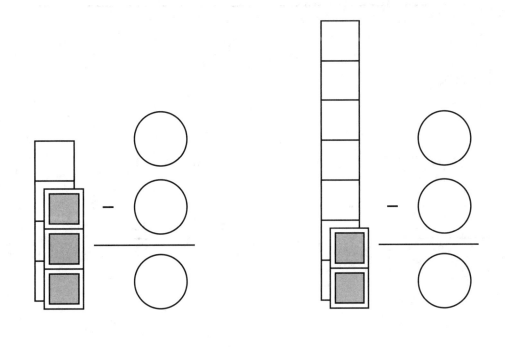

$$\begin{array}{r} 5 \\ -\ 4 \\ \hline \end{array}$$
$$\begin{array}{r} 9 \\ -\ 3 \\ \hline \end{array}$$

$$\begin{array}{r} 4 \\ -\ 2 \\ \hline \end{array}$$

Use tally marks to write the number.

26 = _____

Build and write.

$$
\begin{array}{r}
7 \\
+ 3 \\
\hline
\end{array}
\qquad\qquad
\begin{array}{r}
1 \\
+ 6 \\
\hline
\end{array}
$$

$$
\begin{array}{r}
5 \\
+ 5 \\
\hline
\end{array}
\qquad\qquad
\begin{array}{r}
2\,0\,0 \\
+ 1\,0\,0 \\
\hline
\end{array}
$$

Add.

If the answer is 4, color the space yellow.
If the answer is 6, color the space brown.
If the answer is 8, color the space red
If the answer is 10, color the space orange.

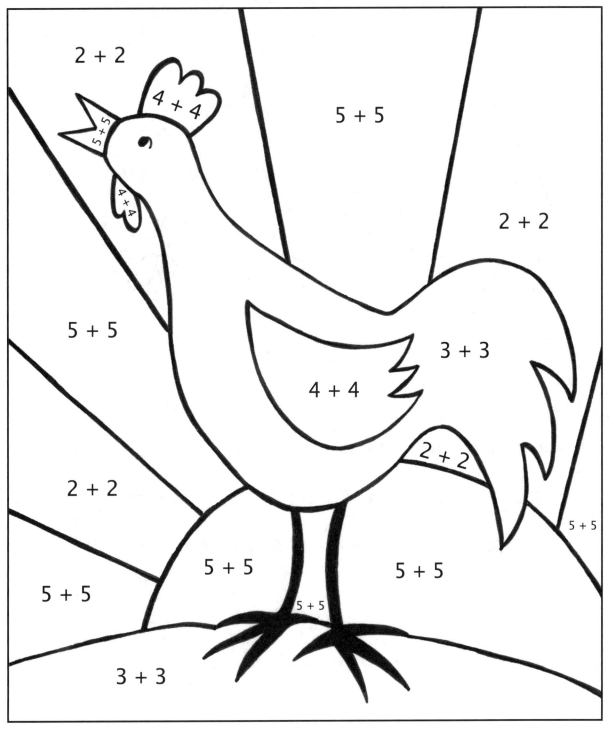

Look at the pictures carefully. Put an X on the one that is different.

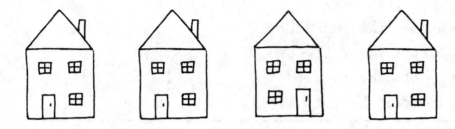

Put an X on the one that is different.

Put an X on the one that is different.

LESSON PRACTICE

Build each problem and write the numbers in the circles.

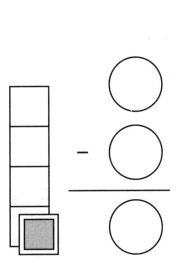

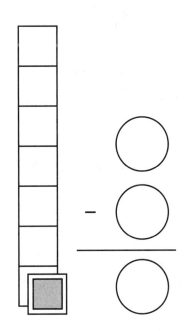

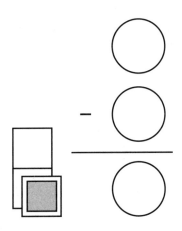

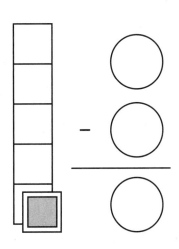

Build each problem. Say and write the answer.

$$
\begin{array}{r} 9 \\ -\ 1 \\ \hline \end{array}
\qquad
\begin{array}{r} 6 \\ -\ 1 \\ \hline \end{array}
$$

30B

Build each problem and write the numbers in the circles.

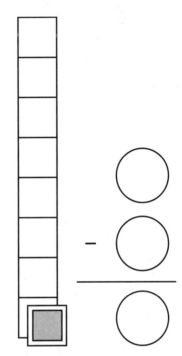

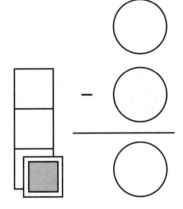

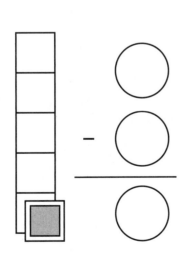

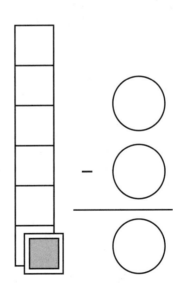

Build each problem. Say and write the answer.

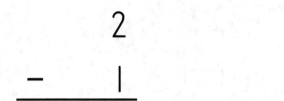

$$\begin{array}{r} 2 \\ -\ 1 \\ \hline \end{array} \qquad \begin{array}{r} 7 \\ -\ 1 \\ \hline \end{array}$$

Build each problem and write the numbers in the circles.

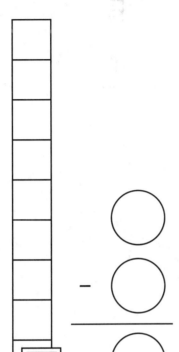

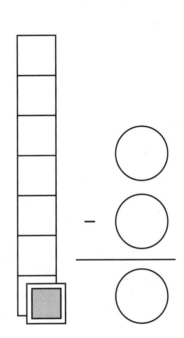

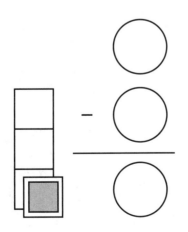

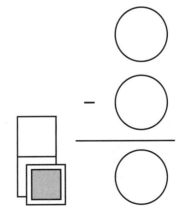

Build each problem. Say and write the answer.

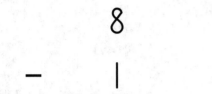

$$\begin{array}{r} 8 \\ -\ 1 \\ \hline \end{array}$$ $$\begin{array}{r} 5 \\ -\ 1 \\ \hline \end{array}$$

Build each problem. Say and write the answer.

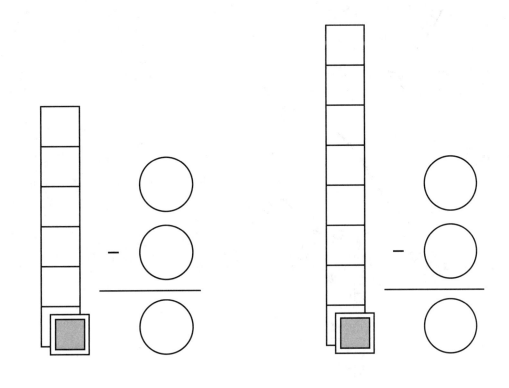

$$\begin{array}{r} 2 \\ -\ 1 \\ \hline \end{array}$$
$$\begin{array}{r} 7 \\ -\ 1 \\ \hline \end{array}$$

$$\begin{array}{r} 3 \\ -\ 1 \\ \hline \end{array}$$

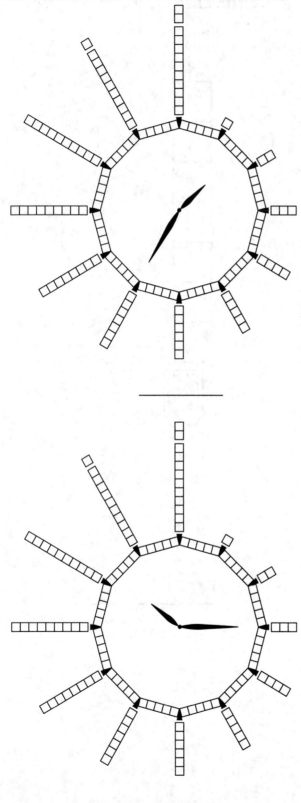

Give the time with hours and minutes.

———————

———————

Build each problem. Say and write the answer.

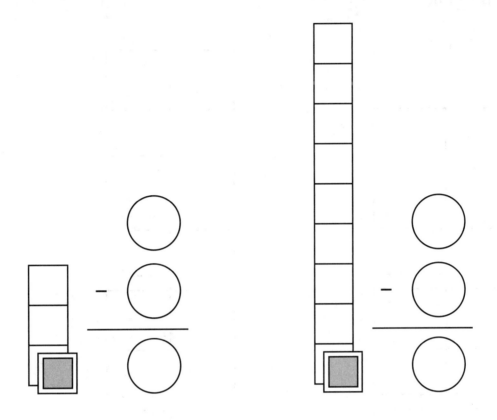

$$\begin{array}{r} 5 \\ -\ 1 \\ \hline \end{array}$$
$$\begin{array}{r} 6 \\ -\ 1 \\ \hline \end{array}$$

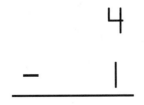

$$\begin{array}{r} 4 \\ -\ 1 \\ \hline \end{array}$$

Find the area of each rectangle by skip counting. Write your
answer in the last box.

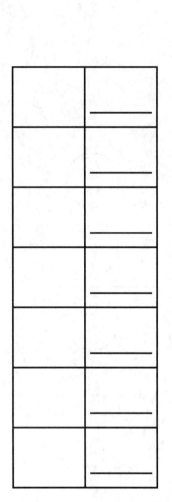

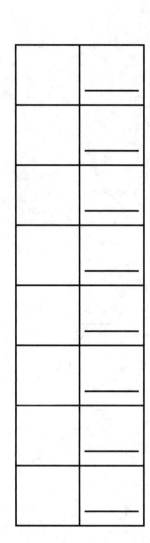

Build each problem. Say and write the answer.

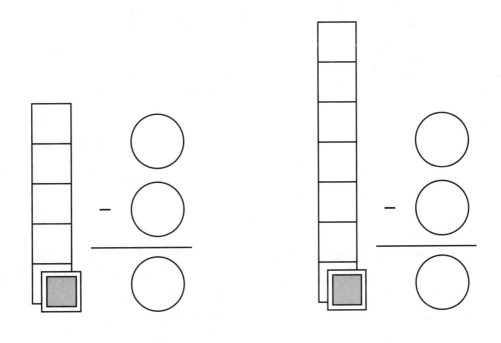

$$\begin{array}{r} 8 \\ -\ 1 \\ \hline \end{array}$$

$$\begin{array}{r} 3 \\ -\ 1 \\ \hline \end{array}$$

$$\begin{array}{r} 2 \\ -\ 1 \\ \hline \end{array}$$

Build and write.

$$
\begin{array}{r}
4 \\
+\ 4 \\
\hline
\end{array}
\qquad
\begin{array}{r}
1 \\
+\ 4 \\
\hline
\end{array}
$$

$$
\begin{array}{r}
8 \\
+\ 2 \\
\hline
\end{array}
\qquad
\begin{array}{r}
30 \\
+\ 30 \\
\hline
\end{array}
$$

$$
\begin{array}{r}
6 \\
+\ 4 \\
\hline
\end{array}
\qquad
\begin{array}{r}
2 \\
+\ 2 \\
\hline
\end{array}
$$

$$
\begin{array}{r}
9 \\
+\ 1 \\
\hline
\end{array}
\qquad
\begin{array}{r}
70 \\
+\ 10 \\
\hline
\end{array}
$$

Draw pictures to help you subtract.

Draw 4 cookies.

$$\begin{array}{r} 4 \\ -\ 1 \\ \hline \end{array}$$

Pat ate one cookie. Put an X on the cookie that Pat ate.

How many cookies are left? _____

Draw 3 trees.

$$\begin{array}{r} 3 \\ -\ 1 \\ \hline \end{array}$$

Dad chopped down one tree.
Put an X on the tree that Dad chopped down.

How many trees are left? _____

Draw pictures to help you subtract.

Draw 5 cars.

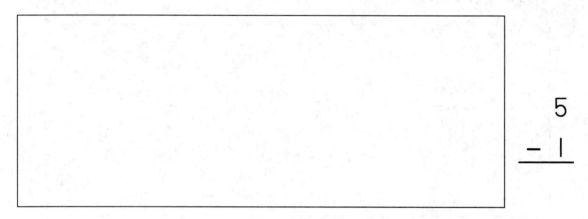

$$\begin{array}{r} 5 \\ -\ 1 \\ \hline \end{array}$$

One car went home. Put an X on the car that went home.

How many cars are left? _____

Draw 6 apples.

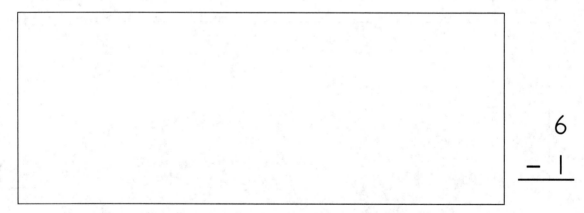

$$\begin{array}{r} 6 \\ -\ 1 \\ \hline \end{array}$$

Mom ate one apple. Put an X on the apple Mom ate.

How many apples are left? _____

CONGRATULATIONS!!

To_____

On this_____ day of _____, 20_____

You have just finished

Primer

You are becoming a math whiz!

Have fun doing the next book, which is

Alpha

Pssssst. Don't forget to thank your teacher!

_____
Instructor's signature